Power Systems

More information about this series at http://www.springer.com/series/4622

J. Raja · P. Ajay-D-Vimal Raj
S. Rajasekar

# Practices in Power System Management in India

Springer

J. Raja
Ministry of Power India
National Power Training Institute
Faridabad
India

S. Rajasekar
Power Grid Corporation of India Limited
Gurgaon
India

P. Ajay-D-Vimal Raj
Department of Electrical and Electronics Engineering
Pondicherry Engineering College
Puducherry
India

ISSN 1612-1287 ISSN 1860-4676 (electronic)
Power Systems
ISBN 978-981-10-2971-4 ISBN 978-981-10-2972-1 (eBook)
https://doi.org/10.1007/978-981-10-2972-1

Library of Congress Control Number: 2018932182



Printed on acid-free paper

This Springer imprint is published by the registered company Springer Nature Singapore Pte Ltd.
part of Springer Nature
The registered company address is: 152 Beach Road, #21-01/04 Gateway East, Singapore 189721, Singapore

*This book is dedicated to our parents and family members*
*and also*
*we wish to thank everyone who helped us to create this book.*

# Preface

This book is recommended for the director, chief engineer, chief accounts officer, superintendent engineer, executive engineer, and assistant engineer who work in Indian distribution companies (DISCOMs).

The book broadly covers the following subjects.

1. In 2008 the government of India announced the Restructured Accelerated Power Development and Reforms Programme (R-APDRP) scheme, a revised version of the Accelerated Power Development Reforms Programme (APDRP). The APDRP scheme was initiated in 2002–2003 as additional central assistance to states in reducing the aggregate technical and commercial (AT&C) losses in the power sector. Aggregate technical and commercial loss captures the total loss in the distribution network. Technical loss may be due to ill-maintained equipment, substations, and inadequate investment in infrastructure whereas commercial loss may be due to low metering efficiency, faulty meter reading, theft and pilferage, and improving the quality and reliability of power supply. This was to be achieved by strengthening and upgrading the subtransmission and distribution system of high-density load centers such as towns and industrial centers.

   Under this scheme Power Finance Corporation (PFC) & Ministry of Power, India (MoP) shortlisted research institutes (RI), and RI published material based on the above area but to date there have been no books covering the basic concept, latest available technology, and practical aspects of operation and maintenance scheduling. This book aims to cover both basics, the latest technology used for distribution and the practical aspects of the operation and maintenance (O&M) of a distribution network. At the end of the book we provide the O&M of all the distribution components with case studies.
2. The Central Electricity Authority of India (CEA) has issued a gazette notification dated September 20th, 2010, whereby they have made it mandatory that all personnel engaged in the O&M of a thermal plant, hydro plant, combined cycle gas plant, transmission system, and subtransmission and distribution system have to be given mandatory training covering the syllabus prescribed by them in this notification.

Presently no study material is available covering the full contents of the syllabus. In order to provide proper training for O&M personnel, it is essential that good study material be available. This book aims to achieve this objective and to meet the requirements of the CEA gazette notification for O&M of the distribution line.

3. A new feature of this book covers the check list of distribution equipment for the O&M of distribution particularly suitable for the Indian power market.

Faridabad, India J. Raja
Puducherry, India P. Ajay-D-Vimal Raj
Gurgaon, India S. Rajasekar

# Acknowledgements

I am deeply indebted to my respected principal director and my business development department members for their invaluable technical support in preparing this book. Finally, I would like to thank my wife, daughters, and my parents for their unconditional love and support. I would not have been able to complete this book without their continuous love and encouragement.

Faridabad, India J. Raja

I would like to express my sincere thanks to the principal of Pondicherry Engineering College, Dr. P. Dananjayan and the head of electrical and electronics engineering, Pondicherry Engineering College, Dr. Alamelu Nachiappan, who have held my hand during the process of learning to carry out this work. Finally, I thank my wife, son, daughter, and my parents for being extremely patient with me in the busy period.

Puducherry, India P. Ajay-D-Vimal Raj

I am thankful to my senior colleagues and fellow engineers at Power Grid Corporation of India Limited for encouraging me to write down my experience which helped me immensely in writing this book. I send my warm thanks to my editors and their teams at Springer, who offered many good suggestions. Finally, I wish to profoundly thank my parents, wife, and family members for all their time, support, and patience.

The views expressed in this book are of the authors only and need not necessarily be those of the organization to which they belong.

Gurgaon, India S. Rajasekar

# Contents

# List of Figures

# List of Tables

# Chapter 1
# Overview of Power Sector Scenario in India

## 1.1 Historical Background and Stages of Growth

The Indian power sector has experienced substantial growth after Independence. The total power generating capacity in India was 1.4 GW at the time of Independence, and as of September 2016 the currently installed capacity was about 306.36 GW (306,360 MW) of which thermal power is 213.22 GW (69.60%), and hydro power 43.11 GW (14.07%); RE power is 44.23 GW (14.44%) and the fastest growing sector among all the sources of power and nuclear power is 5.8 GW (1.89%). Currently, across the globe India has secured the third position in power generation, fourth in consumption, and fifth for installation capacity. The Indian power industry has had significant growth in electricity generation over the decades to 1107.8 BU in 2016, and witnessed 5.6% growth over the previous financial year.

In order to ensure continuous power to all the people, the Indian government has recently announced the launch of the "Power for All" scheme, which emphasizes continuous electricity supply to all domestic industry establishments by building and upgrading necessary infrastructure. Electricity generation was 1091 billion KWh during the year 2015–2016 but had a shortage of 24 billion KWh (lack of −2.1%) against the anticipated −2.2%. The peak load that was generated was 149 GW with a shortfall requirement of 5 GW (lack of −3.3%) against the anticipated −2.7%. In the total world population, nearly 1.4 billon people have no access to electricity, 300 million people of whom are in India. The IEA (International Energy Agency) has found that India has the capacity of adding new generation units of 600–200 GW by 2050. Please see Fig. 1.1. As of September 30th, 2016, nearly 98.5% of 6.0 Lakh villages in India had electricity. However, the supply was intermittent and unreliable in these villages. The Indian government launched *Deen Dayal Upadhyaya Gram Jyoti Yojana* (DDUGJY) as one of its top programs in July 2015, which aims to provide electricity to all rural areas around the clock. They place importance on restructurings and improvement of rural electrification, transmission line strengthening, and the development of distribution systems [1, 2].



J. Raja et al., *Practices in Power System Management in India*,
Power Systems, https://doi.org/10.1007/978-981-10-2972-1_1

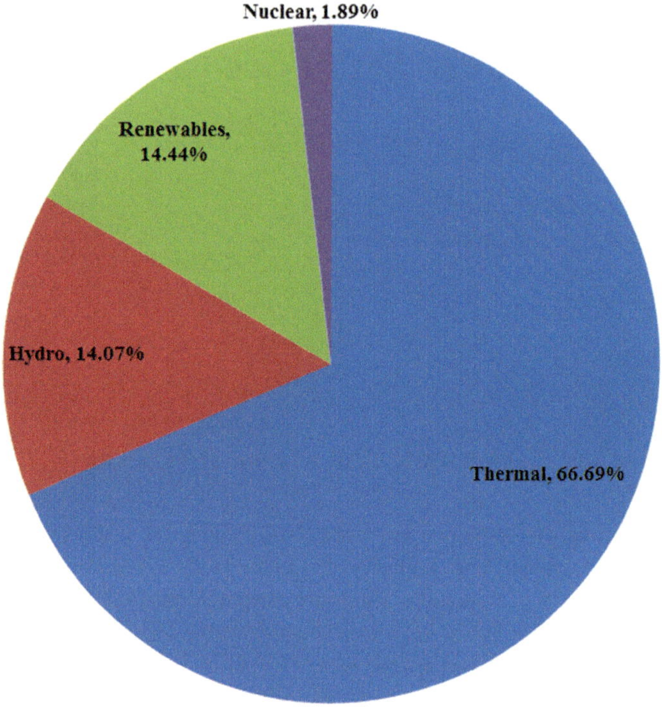


**Fig. 1.1** Percentage of installed capacity for different sources of power (as of September 31st, 2016)

The previous rural electrification scheme, *Rajiv Gandhi Grameen VidyutikaranYojana* (RGGVY), was combined with the DDUGJY scheme.

As a democratic country India relies upon a federal structure. In India's constitution, electricity is a parallel subject: meaning thereby both central and state governments (i.e., parliament and state legislatures) have the power to legislate. The central government prevails over the state government except when the state enacts legislation with presidential assent then (Article 254) State Act will prevail.

The first electric light was seen in India on July 24th, 1879 by a firm called P.W. Fleury & Company at Kolkata, which was a London-based company [3]. After a month, the company changed its name to Calcutta Electric Supply Corporation, however, the control of said company transferred from London to Calcutta only in the year 1970 [4]. After its successful business in Kolkata (Calcutta), it was introduced in Mumbai in 1882 at Crawford market. Later, in 1905, the Bombay Electric Supply & Tramways Company installed a power generating station to provide electricity to tramways. In 1897, India's first hydropower generation was set up in India at Sidrapong by the Darjeeling municipality [5, 6]. In 1905 the first electric street light was introduced in Bangalore.

The first electricity-related legislation was the Indian Electricity Act which inter alia dealt with the basic framework of India's electricity supply, growth, and licenses. However, after Independence, the Electricity (Supply) Act, 1948 was endorsed. This act has facilitated the mandate for establishment of state electricity boards (SEBs) for overall electrification across the country. Furthermore, this 1948 act was amended at various stages [7].

The 1948 Electricity (Supply) Act permitted the formation of SEBs, and was responsible for generation, and transmission & distribution (T & D) of power within the state. Later the Central Electricity Authority (CEA) was established to administer the planning and development of the power sector and guide SEBs. Electricity was placed under the concurrent authority of central and state governments in the Indian constitution.

In 1975, revisions were made to enable the central government to set up and maintain power plants. Consequently, the National Thermal Power Corporation (NTPC) was formed which is the largest power generation company in India to date. Subsequently, different power companies including NHPC (1975), NEEPCO (1976), NPCIL (1987), and THDC (1988) were formed and owned by the central government to deal with the generation of power mainly in the thermal, hydro, and nuclear sectors.

The summary of events is as follows.

- **Year 1975**: Incorporate provisions to enable generation in the central sector (NTPC, NHPC, etc.). Incorporate provisions to facilitate state electricity boards (SEBs) for earning minimum returns.
- **Year 1991**: Permit private sector participation in generation. Also formation of regional load dispatch centers (RLDCs).
- **Year 1998**: Permit private sector participation in the transmission sector. Also the establishment relating to State Transmission Utility (STU) and Central Transmission Utility (CTU).

Apart from the above modifications in the earlier act, the major milestone was the Electricity Regulatory Commission Act, 1998 which had significant facility for setting up of a regulatory commission at central and state levels.

- **In the power sector, the country was classified into five regions, namely**: (i) Northern Region (NR), (ii) Eastern Region (ER), (iii) Southern Region (SR), (iv) Western Region (WR), and (v) North Eastern Region (NER). Each region contains several states (or state utilities); for example: WR contains utilities of Madhya Pradesh, Gujarat, Maharashtra, Chhattisgarh, Dadara and Nagar Haveli, Goa, and so on.

## 1.2 Major Players in the Indian Power Sector

(i) **National Thermal Power Corporation (NTPC)**: NTPC produces the highest amount of power in India, with installed plant capacity of 47 GW including joint ventures. In the near future, NTPC aims to reach 126 GW of power generation capacity.

(ii) **TATA Power**: The largest integrated power company in India is TATA Power, with a substantial presence in the renewable energy sector. The company owns more than half the private sector total power generation. It had 10 GW generating capacities in fiscal year 2016. Over the next six years the company plans 18 GW generation units and also wants to increase its distribution network by 4 GW and energy sources by 25 million tons per year.

(iii) **Reliance Power**: The company has about 35 GW of power capacity, including operational (6 GW) and proposed.

(iv) **CESC Limited**: CESC Limited works in coal mining and generation, as well as the distribution sector of the power system. With complete control of three coal-based power plants together they generate 1.3 GW.

(v) **NHPC**: It is the biggest hydroelectric utility in India, with a production capacity of 6.5 GW; it added 6.7 GW of power in 2017. NHPC is constructing nine projects, aggregating an installed capacity of 4.2 GW.

(vi) **Power Finance Corporation Limited (PFC)**: PFC provides financial support to power utility companies for their power project developments. Their major services include assisting short-term loans, project-term loads, financing, and consultancies.

(vii) **Adani Power**: The total power generation capacity of Adani Power in 2016 was 10.5 GW and by 2020 they plan to increase their generation capacity to 20 GW. Now they are one of the largest thermal power producers and their thermal power plant in Mundra, Gujarat is considered the world's largest thermal power plant.

(viii) **PGCIL**: The Power Grid Corporation of India Limited is one of the world's largest transmission utility companies. They are accountable for planning interstate transmission systems, co-ordination, and implementing new transmission networks in India. In their twelfth five-year plan PGCIL aims to build 72.25 GW capacity transmission networking; in 2016 their network capacity was 47.45 GW.

(ix) **DVC**: Damodar Valley Corporation is responsible for developing the power sector in irrigation and flood control systems.

(x) **SJVNL**: SJVNL is a joint venture between the government of India and the government of Himachal Pradesh; their main focus is to develop hydropower generation in India. Now they are expanding their focus area on wind power projects.

## 1.3 The Growth Story of the Indian Power Sector

India is the third largest producer of electricity in the world. Generating capacity has grown from 1.4 GW in 1947 to 306 GW in 2016 (Tables 1.1, 1.2, and 1.3). Overall generation in India has increased from 301 billion units (BUs) during 1992–1993 to 1107.8 BUs in 2015–2016. The country saw 5.64% growth over the year 2015–2016.

**Table 1.1** Growth of electricity installed capacity in India

| As of/during financial year ending with | Installed capacity (MW) | Per capita consumption (kWh) |
|---|---|---|
| 31-Dec-1947 | 1362 | 16.8 |
| 31-Dec-1950 | 1713 | 18.2 |
| 31-Mar-1956 | 2886 | 30.9 |
| 31-Mar-1961 | 4653 | 45.9 |
| 31-Mar-1966 | 9027 | 73.9 |
| 31-Mar-1969 | 12,957 | 97.9 |
| 31-Mar-1974 | 16,664 | 126.2 |
| 31-Mar-1979 | 26,680 | 171.6 |
| 31-Mar-1980 | 28,448 | 172.4 |
| 31-Mar-1985 | 42,585 | 228.7 |
| 31-Mar-1990 | 63,636 | 329.2 |
| 31-Mar-1992 | 69,065 | 347.5 |
| 31-Mar-1997 | 85,795 | 464.6 |
| 31-Mar-2002 | 105,046 | 559.2 |
| 31-Mar-2007 | 132,329 | 671.9 |
| 31-Mar-2008 | 143,061 | 717.1 |
| 31-Mar-2009 | 147,965 | 733.5 |
| 31-Mar-2010 | 159,398 | 778.6 |
| 31-Mar-2011 | 173,626 | 813.3 |
| 31-Mar-2012 | 199,877 | 883.6 |
| 31-Mar-2015 | 271,722 | 1010.0 |
| 31-Mar-2016 | 301,965 | 1250.2 |
| 30-Nov-2016 | 306,360 | 1265.3 |

**Table 1.2** PAN India power supply situation in FY 2016–2017

| Energy | | |
|---|---|---|
| Region | Requirement (MU) | Availability (MU) |
| Northern | 3,57,459 | 3,51,009 |
| Western | 3,79,087 | 4,05,370 |
| Southern | 3,10,564 | 3,20,944 |
| Eastern | 1,51,336 | 1,35,713 |
| Northeastern | 16,197 | 14,858 |

During fiscal years 2010–2016, India's electricity generation grew at the compound annual growth rate (CAGR) of 6.21%. The twelfth five-year plan predicts a domestic energy generation of 669.6 MTOE (million tons of oil equivalent) rise to 844 MTOE by the year 2021–2022.

Table 1.3 Sectorwise utility power generation capacity as of November 2016

| | | Central | State | Private | All India | % |
|---|---|---|---|---|---|---|
| Thermal (MW) | Coal | 51390.00 | 64210.50 | 70992.38 | 186592.88 | 61.06 |
| | Gas | 7490.83 | 7210.70 | 10355.60 | 25057.13 | 8.20 |
| | Diesel | 0.0 | 363.93 | 554.96 | 918.89 | 0.03 |
| | Subtotal thermal | 58880.83 | 71785.13 | 81902.94 | 212568.90 | 69.56 |
| | Nuclear | 5780.00 | 0.0 | 0.0 | 5780.00 | 1.89 |
| Renewable (MW) | Hydro | 11651.43 | 28197.00 | 3120.00 | 42968.43 | 14.06 |
| | Other renewable | 0.0 | 1963.80 | 42273.12 | 44236.92 | 14.47 |
| | Subtotal renewable | 11651.43 | 30055.81 | 44236.92 | 87205.35 | 28.54 |
| Total (MW) | | 76312.26 | 101945.93 | 127296.06 | 305554.25 | 100 |

## 1.4 The Growth in Rural Electrification

During Independence more than 80% of the population lived in rural India. Actual development of the country can only be possible with the development of the rural sector. The government has introduced various plans to improve the standard of living in villages with the focus on rural electrification (Table 1.4). As of September 30th, 2016, nearly 98.5% of 6.0 Lakh villages in India were electrified.

## 1.5 Growth in Generating Capacity Addition

Table 1.5 shows the addition in generating capabilities of the post-Independence power sector in India.

Table 1.4 Rural electrification status in India

| States and Union Territory | Number States and UT | Rural electrification rates (%) |
|---|---|---|
| Andra Pradesh, Haryana, Kerala, Goa, Maharashtra, Punjab, Sikkim, T. Nadu, Telengana, Chandigarh, Dadra, Delhi, Damman and Diu, Lakshadweep, Puducherry | 16 | 100 |
| Karnataka, UttaraKhand, Himachal Pradesh, Madhya Pradesh, West Bengal | 5 | 99 |
| Bihar, Rajesthan, Chhattisgargh, Jammu and Kashmir, Tripura, Uttar Pradesh | 6 | 95 |
| Jharkhand, Odisha, Nagaland, Mizoram | 5 | 80 |
| Arunachal Pradesh | 1 | Under 80 |

**Table 1.5** Growth of installed generation capacity in India

| As of/during financial year ending | Hydro | Thermal | | | Total Thermal | Nuclear | RES* | Total |
|---|---|---|---|---|---|---|---|---|
| | | Coal/ lignite | Gas | Diesel | | | | |
| 31-Dec-1947 | 508 | 756 | 0 | 98 | 854 | 0 | 0 | 1362 |
| 31-Dec-1950 | 560 | 1004 | 0 | 149 | 1153 | 0 | 0 | 1713 |
| 31-Mar-1956 | 1061 | 1597 | 0 | 228 | 1825 | 0 | 0 | 2886 |
| 31-Mar-1961 | 1917 | 2436 | 0 | 300 | 2736 | 0 | 0 | 4653 |
| 31-Mar-1966 | 4124 | 4417 | 134 | 352 | 4903 | 0 | 0 | 9027 |
| 31-Mar-1969 | 5907 | 6640 | 134 | 276 | 7050 | 0 | 0 | 12,957 |
| 31-Mar-1974 | 6966 | 8652 | 165 | 241 | 9058 | 640 | 0 | 16,664 |
| 31-Mar-1979 | 10,833 | 14,875 | 168 | 164 | 15,207 | 640 | 0 | 26,680 |
| 31-Mar-1980 | 11,384 | 15,991 | 268 | 165 | 16,424 | 640 | 0 | 28,448 |
| 31-Mar-1985 | 14,460 | 26,311 | 542 | 177 | 27,030 | 1095 | 0 | 42,585 |
| 31-Mar-1990 | 18,307 | 41,236 | 2343 | 165 | 43,744 | 1565 | 0 | 63,616 |
| 31-Mar-1992 | 19,194 | 44,791 | 3095 | 168 | 48,054 | 1785 | 32 | 69,065 |
| 31-Mar-1997 | 21,658 | 54,154 | 6562 | 294 | 61,010 | 2225 | 902 | 85,795 |
| 31-Mar-2002 | 26,269 | 62,131 | 11,163 | 1135 | 74,429 | 2720 | 1628 | 105,046 |
| 31-Mar-2007 | 34,654 | 71,121 | 13,692 | 1202 | 86,015 | 3900 | 7760 | 132,329 |
| 31-Mar-2008 | 35,909 | 76,049 | 14,656 | 1202 | 91,907 | 4120 | 11,125 | 143,061 |
| 31-Mar-2009 | 36,846 | 77,649 | 14,876 | 1200 | 93,725 | 4120 | 13,242 | 147,933 |
| 31-Mar-2010 | 36,863 | 84,198 | 17,056 | 1200 | 102,454 | 4560 | 15,521 | 159,398 |
| 31-Mar-2011 | 37,567 | 93,918 | 17,706 | 1200 | 112,824 | 4780 | 18,455 | 173,626 |
| 31-Mar-2012 | 38,990 | 112,022 | 18,381 | 1200 | 131,603 | 4780 | 24,504 | 199,877 |
| 31-Mar-2013 | 39,491 | 130,221 | 20,110 | 1200 | 151,531 | 4780 | 27,542 | 223,344 |
| 31-Mar-2014 | 40,532 | 145,273 | 21,782 | 1200 | 168,255 | 4780 | 31,692 | 245,259 |
| 31-Mar-2015 | 41,267 | 164,636 | 23,062 | 1200 | 188,898 | 5780 | 35,777 | 271,722 |

*Renewable energy sources also include hydro capacity of 25.0 MW and below as reported by MNRE

## 1.6 Electricity Generation in India

During the early years after Independence, India's electricity generation was very low compared to other developed nations. But after 1990, faster growth was recorded in the electricity generation sector. For example, electricity generation increased from 179 to 1279 TW-hr in 1985 and 2015, respectively. The contribution of coal-fired plants and nonconventional energy sources are major components of the total energy generation growth; however, the contribution of fossil fuels and hydro has significantly decreased in the period 2011–2016. During the fiscal year 2016–2017, India has as its target the generatation of 1173 billion kWh which excludes all types of renewables, compared to the equivalent generation of 1.1 thousand billion kWh.

### 1.6.1 *Thermal Power*

In India coal power contributes about 70% of the total electricity produced in the country. The Indian government has mandated the use of coal with ash content less than 34% in plants near urban and biologically sensitive areas. For this reason, the coal support industry has grown significantly, with a current capacity of 90 MT. As per the latest emission norms the investment of INR 12.5 million per MW is required for installation of pollution control devices in a thermal power plant.

**Coal Supply Constraint**
Indian coal reserves have low carbon content, low calorific value, and negligible toxic traces. The natural fuel value of Indian coal is very poor and consumes about 700 gms of coal per p1 kWh with a gross calorific value (GCV) of 4550 kcal/kg, but the quality of coal from other parts of the world is much better. The Indian thermal power plant sector imports nearly 95 metric ton of high steam-grade coal which contributes nearly 29% of the cumulative consumption in India to meet the demand in steel production and other thermal power plants.

**Oil and Gas**
The policy debate for oil and gas is usually made jointly due to various co-relationships such as the carbon-to-heat ratio content and the easy, low cost accessibility of gas. At present the global shares of oil and gas contributing to energy consumption are 31 and 22%, respectively; per the IEA's prediction they would be 25% by 2035. Oil share demand is slowly coming down whereas gas demand is increasing. Meanwhile in 2016, the global LNG has been dipping radically because of excess production but the landed costs are inexpensive for coal-based electricity generation, due to the lack of adequate degasification capacity of LNG.

**Changing Old Thermal Power Plants**
Most of the thermal power plants existing in India are inefficient and need to adapt the latest technology in order to reduce greenhouse gas emissions. Each of the existing thermal power plants in India emits nearly 50–120% higher pollutants as compared to thermal plants existing in developed countries. To reduce higher emissions the Indian government invested nearly 70,000 crore to shut down nearly 11 GW existing facilities and replace them with the latest supercritical technology totaling more than 20 GW. Furthermore, over 100 outdated plants will be replaced by 30 supercritical plants of 660 MW to 800 MW, thus saving Rs. 40,000 crore.

### 1.6.2 *Renewable Energy*

As of March 2017, India had a generation capacity of approximately 60 GW based on renewable technologies, constituting about 15.5%, and exceeding the total hydroelectric power capacity of India. Due to global warming, the government is promoting the use of RE energies; also the recent maturity of RE technologies and cost reduction motivates RE adoption at a faster pace. By 2022 a renewable energy system addition in the power system network is targeted at 175 GW and in 2040 it is likely to be 624–850 GW. This will translate into 50–60% installed capacity and 31–41% RE generation by 2030, in place of 13 and 7%, respectively, in 2014–2015.

### *1.6.3 Hydro*

In 1898 and 1902 at Darjeeling and Shivana samudra some hydropower plants were established. India is blessed with hydro potential of about 84,000 MW at 60% load factor. Further to this, 6740 MW from small, mini, and micro schemes have been estimated and 94 GW capacities and 56 sites for pumped storage schemes have been identified.

### *1.6.4 Nuclear Power*

Up until April 2016, India had 5.78 GW; in other words 1.91% of total power generation capacity and an ambitious plan is underway to ramp up this source by 2022 with six reactors under construction having a cumulative capacity of 4300 MWe, and six more planned with a capacity of 4800 MWe. Once the above capacities are realized, India's installed capacity for nuclear power is likely to go up substantially. Additionally, seven sites have received in principle approval for building new reactors. The above capacity enhancement agenda will give a solid foundation to this sustainable source of electricity in India's power development.

## 1.7 Transmission and Distribution

The length of India's HV transmission lines (700,000 km) is comparable to that of the United States, however, with far less capacity. As of March 31st, 2015 the total length of 400 V lines (excluding 220 V) was around $10.56 \times 10^6$ km distance. The 400 V transmission lines can form a square matrix with an area 37 $km^2$ of the country and the peak loading should not exceed more than 153 GW capacity; thus far the peak load was met around 148 GW. The achieved maximum demand factor of each substation does not exceed 60% at the 200 kV level and the aggregate technical & commercial losses (AT&C) are around 27%. The government has fixed the national AT&C loss target at around 24% for the year 2011 and further, reducing them to around 17.1% by 2017 and to 14.1% by 2022.

## 1.8 Unbundling of Power Sector

The planning of the transmission system in India is linked with new generation projects as part of the evacuation system. The central sector generators such as NTPC and NHPC built their own transmission networks for the transmission of power to different parts of India. Later, the government integrated the central and state transmission systems and formed a national power grid to facilitate transmission of central sector power generation. These efforts led to the incorporation of a separate entity known as the National Power Transmission Corporation in 1989 which was later renamed POWERGRID. This new company POWERGRID was born with the amalgamation of total transmission assets and personnel of various generating companies including NTPC, NHPC, NJPC, NEEPCO, NLC, and THDC. Thus, with this a new era began where a separate company, POWERGRID, started its business of interstate transmission of power by the central sector power generating companies. Please see Fig. 1.2.

The central sector transmission system was separated from generation as an outcome of power sector reforms and a strong central transmission network of 96,000 circuit kms with around 160 substations (EHVAC/HVDC) having a conversion capacity of 138,673 MVA and interregional power transfer capacity of 28,000 MW was created within a span of two decades by POWERGRID which is also discharging its role as a central transmission utility (CTU).

The CTU is essential in interstate transmission network planning, co-ordination, and expansion at the local as well as national level. Eleven high-capacity transmission corridors mainly comprising state-of-the-art 765 DC and ±800 kV

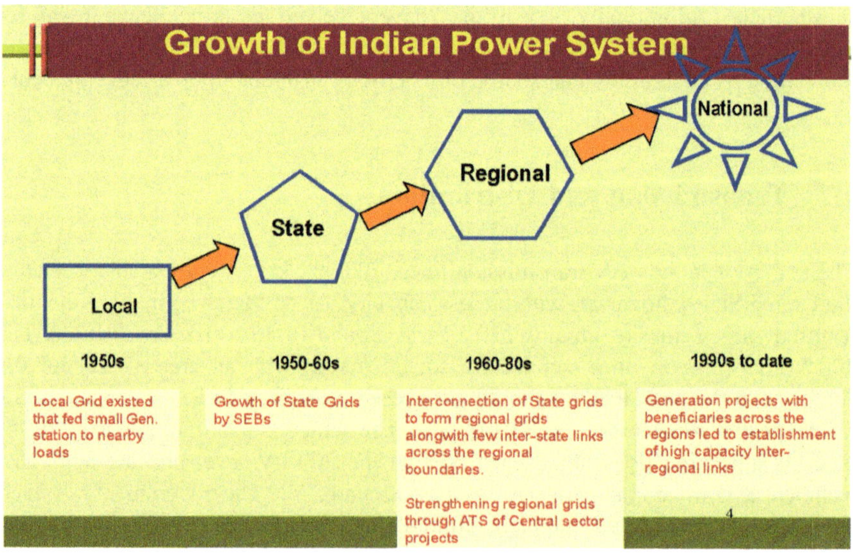


**Fig. 1.2** Growth in Indian power sector

6000 MW have been developed in the national grid to enhance transmission line capacity to facilitate bulk-load centers. Furthermore, to encourage the additional capacity program, the CTU is providing connectivity, medium-term open access (MTOA), and long-term access (LTOA) to various generation developers as per CERC regulations. In view of the need for defensible growth, the CTU is focused towards developing technologies such as UHVAC, high-voltage DC, FACTS, smart grids, and the like in transmission development, which is unique in the world.

The Indian power system is growing manifold along with complexity in transmission system operation, posing challenges in maintaining grid security, reliability, and stability. To address these issues, it is prudent to introduce transmission intelligence through smart grid applications. In this direction, POWERGRID, the CTU, has undertaken full-scale implementation of state-of-the-art synchrophasor technology using phasor measurement units (PMU) at all 400 kV and above substations and 220 kV generation switchyards in the country for a wide area measurement/monitoring system (WAMS) of the power network integrating state and central grids in a unified manner. During the operation stage, the CTU carries the finance of ISTS transmission costs in addition to signing TSA and RSA. Tariff-based competitive bidding (TBCB) has been in effect from January 6th, 2011 for transmission projects and is also valid for the CTU. In this emerging environment of competition, POWERGRID has successfully forayed into the competitive scenario by bagging both transmission projects floated under the TBCB route.

## 1.9 Private Participation in Transmission

(i) **Joint Ventures**:

Due to its capital-intensive nature, huge investments are required in the transmission sector to keep pace with the power sector growth as it meets the ever-growing demand for power. Presently transmission is primarily owned by the public sector and public funds are not sufficient to meet this requirement. Therefore large investments are also required from the private sector for additional funds to supplement public resources in order to keep pace with the power sector expansion. To attract private investment, the Indian government has provided various regulatory and legal frameworks. The enactment of the Electricity Laws (Amendment) Act 1998 and Electricity Regulatory Commission Act 1998 has facilitated the formation of the Central Electricity Regulatory Commission (CERC), State Electricity Regulatory Commissions (SERCs), Central Transmission Utility (CTU), and State Transmission Utilities (STUs) and has also paved the way for entry of private sector participation in the transmission business as "transmission licensees." In order to infuse resources from the private sector, the Indian Ministry of Power introduced guidelines for private sector participation in the transmission sector (vide ref No. 9/3//98-PG dated 31.1.2000). These courses of action provide two different options for private sector participation in transmission.

(i) A joint venture (JV) should be selected through the competitive bidding process called the JV route. The CTU/STU should contribute a minimum of 26% equity and the balance contributed by the JV partner.
(ii) The responsibility of the CTU/STU will identify permitted transmission to be implemented by the private entity, also facilitate infusion of resources from the private sector. The Independent Power Transmission company (IPTC) is an implementing agency where their 100% equity is owned by the private sector.

From the above, it can be seen that suitable policy exists to promote private sector participation in transmission. A number of independent power producers can establish their own projects in various parts of the country. They need to inject part and/or all of the power generated to the national grid owned by POWERGRID. For injecting power generated from their generating project to the POWERGRID national grid, a dedicated transmission system is required to be constructed connecting the generating project to this national grid. Per the Electricity Act, 2003 (Sect. 10), it is the duty of the generating company to establish a dedicated transmission system. Because many generating companies do not have the expertise to establish a transmission system, they make a request to POWERGRID to become a 26% equity partner in a JV company to implement such dedicated transmission systems. Note that the tariffs for JVs are computed on a cost-plus basis and guided by CERC tariff regulations issued by them from time to time.

(ii) **IPTC Route**:

The second route for private sector participation is through IPTC, wherein a private entity owns 100%. Many transmission systems in the past have been routed through the IPTC process and finally the government of India issued an order regarding development of interstate transmission systems on the tariff bidding route from January 5th, 2011 to encourage competition. Earlier the transmission projects were executed on the cost-plus basis. With the advent of private players and more complexity in the sector, it was necessary to introduce tariff-based competitive bidding. These guidelines allow private players to enter the transmission sector through broader contributions in providing transmission services, thereby facilitating competition. It helps in transparently awarding the project based on the formalized tariff.

## 1.10 Electricity Act and Policy

It is a multifaceted or complex Act purporting to combine the laws relating to the generation, transmission, distribution, trading, and use of electrical energy and for conducive measures for the development of the power industry. It replaces the three previous pieces of legislation:

(i) The Indian Electricity Act, 1910
(ii) The Electricity (Supply) Act, 1948
(iii) The Electricity Regulatory Commissions Act, 1998.

The main aim of this act was electricity supply to all areas, consumer protection, electricity tariff realization, transparent policy regarding subsidies, promoting efficient and environmentally benign policies, and the constitution of the CEA, CERC, and SERCs and establishment of an appellate tribunal [8]. Provisions for open access in transmission have also been considered wherein the transmission systems of a transmission licensee are to be made available for nondiscriminatory open access. On the policy front, the central government has been made responsible for preparing, publishing, and revising the national electricity policy (NEP) and tariff policy in consultation with the states. The tariff policy has always been a matter of concern and has led to many litigation cases and chaos [9, 10]. With new amendments in the act the responsibility of formulating tariff policy falls on the central government and CERC.

### *1.10.1 Provisions in the Electricity Act*

The Electricity Act 2003 (Fig. 1.3) covers all the branches relating to the power sector. Some of its key areas and the points they cover are:

- **Generation**:

  (a) De-licensing
  (b) Liberal provisions in captive generation
  (c) Rural generation freed from licensing

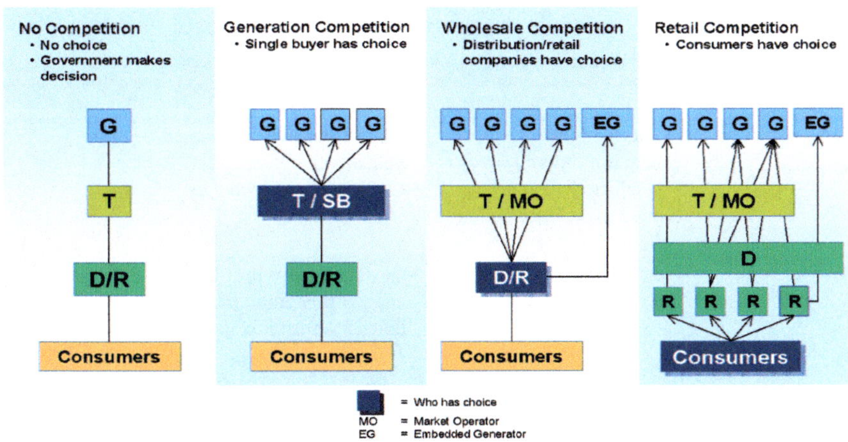


**Fig. 1.3** Implementation of Electricity Act

- **Transmission and Distribution**:
  (a) Open access in transmission/distribution systems
  (b) Envisages unbundling of transmission and distribution
  (c) Rural distribution freed from licensing
  (d) Provision for issuing more than one license for distribution within the same area
- **Trading and Markets**:
  (a) Trading in electricity permitted
  (b) Regulatory commissions to develop electricity markets.
- **Other Features**:
  (a) Expanded role for the regulatory commissions
  (b) Provision for setting up an appellate tribunal for disposal of appeals against CERC/SERC
  (c) Provisions for preventing and eliminating power theft and stringent penalties for the latter.

A new industry structure has evolved after the implementation of the Electricity Act 2003, which can be summarized as the following.

### *1.10.2 Impact of Electricity Act*

It can thus be seen that the Electricity Act 2003 (Tables 1.6 and 1.7) has been a very important instrument in the power sector wherein each and every aspect has been considered and the proper direction for each matter has been tried to be addressed. In this background, the details of various acts that had underlain the power sector in India till now are described below.

**Table 1.6** Impact of the Electricity Act 2003 in different segments of electricity

| Segment | Objective | Impact |
|---|---|---|
| Generation | • De-licensing the generation units<br>• Liberalization in the captive power generation policy | • More companies motivated towards generation<br>• Captive power increases |
| Transmission | • Fair access to T and D lines | • Customer can select by their own choice/ competition amongst DISCOMs leads to efficient transfer of power |
| Distribution | • Fair access in phase manner<br>• Strict punishments for power theft<br>• Lucid subsidy management | • Choice for buyer to choose supplier<br>• Loss reduction<br>• Equal benefit to all |

**Table 1.7** Electricity Act, objective and impact

| Laws/Policies | Objective | Impact |
|---|---|---|
| The Electricity Act, 1910 | Infrastructural framework for supply of electricity | Attracted private capital |
| The Electricity Act, 1948 | Mandated creation of SEBs | Ownership in the hands of SEBs |
| IIP Process, 1991 | Private investment in power generation | Generated projects from private players |
| The Electricity (Amendment) Act, 1998 | Making transmission a separate activity | Central transmission utility and state transmission utilities set up |
| Mega power policy, 1995 | Setting up of ultra mega power plants | Mega power plants benefit |
| The Regulatory Commission Act, 1998 | Provision for setting up of central/ state electricity regulatory commission | Independent regulatory mechanism |
| National Electricity Policy | Competition and protection of consumer | More players influenced to invest and more efficient consumer service |
| Electricity Act, 2003 | Providing reliable and quality power to consumers at reasonable rate | Investment in capacity addition |
| National tariff policy | Tariff structuring | Attractive tariff for players |

- **Generation**:

  (a), (b) Liberal provisions in captive generation
  (c) Rural generation freed from licensing

- **Transmission and Distribution**:

  (a) Open access in transmission/distribution systems
  (b) Envisages unbundling of transmission and distribution
  (c) Rural distribution freed from licensing
  (d) Provision for issuing more than one license for distribution within the same area.

## 1.11 Various Policies

(i) **National Electricity Policy (NEP) and National Tariff Policy (NTP)**

**Part II of the Electricity Act mandates that** the central government prepare the NEP and NTP with the state governments of India, and other authorities responsible for the growth of power system fossil resources, hydro, and renewable sources of energy.

(ii) **National Electricity Policy (NEP)**

The National Electricity Policy has been developed in discussion with and taking into consideration state governments, the central electricity regulatory commission (CERC), central electricity authority (CEA), and other stakeholders [11, 12]. The key objective of NEP is specifying guidelines for speedy development of the power sector.

(iii) **Aims and Objectives**

- Provide electricity to all families in the forthcoming five years.
- Aim to meet the power demand and have reasonable reserve available.
- Provide reliable high-quality power at nominal rates.
- Aim to increase per capita electricity.
- Financial turnaround and commercial feasibility of electricity sector.
- Protection of consumers' interests.

(iv) **National Electricity Plan**

The CEA is responsible for short-term planning, whereas NEP would be for a short-term framework of five years while giving a 15-year viewpoint and would include:

- Collects demand forecasts for different regions in short- and long-term perspectives
- Recommends capacity addition in view of generation and transmission, regarding economics, system losses, requirement of load centers, grid stability, supply security, power quality including voltage, frequency profile, and the like, and environmental considerations including rehabilitation and resettlement
- Integration of such possible locations with T & D of national grid including requirement of redundancies.

(v) **Issues Addressed in the NEP**

- Rural electrification and power generation
- Transmission and distribution
- Recovery of cost of services and targeted subsidies
- Technology development and research and development (R&D)
- Consumer benefits from competition
- Financing power sector program including private sector participation
- Energy conservation and other environmental issues
- Training and HR development
- Cogeneration and NCES.

(vi) **National Tariff Policy, 2006**

In compliance with Section 3 of the Electricity Act, the central government noted the tariff policy subsequent to the NEP. The new tariff policy is for ideal growth of the electricity transmission network to endorse effective consumption of generation and transmission resources in the country and appealing for needed investments in the said sector with satisfactory returns. A new policy also ensures the availability of power at nominal rates and addresses viability and attracts investments.

(vii) **Objectives of the Implementation of the NEP and Tariff Policy**

- Optimal development of the transmission network.
- A suitable transmission tariff framework for encouraging effective utilization of assets and aiming for faster growth of new transmission networks.
- Transmission charges are on the basis of MW/circuit/km, on the zonal postage stamp basis, and/or some other realistic variant.
- To share the total transmission cost in proportion to respective utilization of the transmission system.
- In view of the method directed by the NEP, prior agreement with the legatees would not be a precondition for network expansion.
- In a period of one year, the CERC is to establish standards for capital and operating costs, operating standards, and performance indicators for different types of transmission lines.
- The tariff of the projects to be developed by the CTU/STU would also be determined on the basis of competitive bidding after five years or when the regulatory commission is satisfied that the situation is right to introduce such competition.
- Transactions are charged on the basis of average losses after appropriately considering the distance and sensitivity (directional), as applicable to the relevant voltage level.
- The loss compensation to be reasonable and linked to applicable technical loss benchmarks.
- The CERC should permit satisfactory capital investments in new assets for upgrading the transmission system.

## 1.12 National Energy Policy

The key objectives of Indian energy policies are (i) energy access at a reasonable price, (ii) improved security, (iii) greater security, and (iv) economic growth.

(i) **Energy access at reasonable price**:

It is essential to provide energy access at reasonable cost particularly for the rural population in India. Nearly 304 million people do not have access to electricity for daily needs. 30% of the Indian population is housed and are still dependent upon biomass. By 2022, the government of India (GoI) has targeted the electrification of every household as per budget planning for 2015–2016.

(ii) **Improved security**:

The policy also includes providing energy security by decreasing reliance on conventional energy resources. Presently, India greatly relies on oil and coal imports while also importing some gas. Energy security could be compromised in the case of any interruption of imports, and therefore the focus on domestic energy production is increased to meet the demand [13]. To complement the diversification of energy resources, the GoI is also planning to make agreements regarding import supplies.

(iii) **Greater security**:

The influence of climate change and deterioration of air quality owing to fossil fuel consumption has changed the perception of sustainability goals [14]. As sustainability is also dependent on energy security, the current scenario of importing 90% of the energy supply has to be adapted to focus on renewable energy production.

(iv) **Economic growth**:

Sustainable and fast economic growth is the energy policy objective of India. By promoting growth through efficient energy resources, there are two benefits.

1. Energy is identified as the key factor of economic development and its availability at affordable cost is important to energy-intense infrastructure.
2. Because energy production is itself identified as a huge market, its growth will have an impact on overall economic growth.

Energy is identified as a critical resource to improve the living standards of societies across the country based on the correlation between energy consumption and the human development index (HDI). Inasmuch as the overall interest of the Indian government is to eradicate poverty, to achieve it the country has revised its energy policies over the years to increase per capita electricity consumption. With India accommodating 30% of the world population and almost 304 million without access to energy and relying on solid biomass for household needs, it is a challenge. India is attempting to realize a two-digit growth rate with respect to its national income, and energy being an important sector, producing clean energy to all households has to be a major factor of the poverty eradication program.

To fulfill GoI proposal plans on energy development, the national energy policy will draft their policies. The top priority goal of GoI is to electrify all villages that don't have electricity access by 2022. The carbon footprint is targeted to be reduced by 35% in 2030; this could be achieved by promoting renewable energy power generation. Accordingly the NEP creates the energy policy for the country's energy usage and development plans.

## 1.13 Energy Demand: Efficiency and Conservation

India's share of the world population is 17%; conversely the shares in the world fossil reserves are only 1.5% combined. There is high dependence on fossil fuel imports even at such a low level of consumption. Recent economizing policy has focused on demand-side interventions on energy usage. One of the said interventions is behavioral change that results in reducing the demand for energy-based service, called energy conservation; the second intervention is greater energy efficiency [15, 16]. To expound energy conservation let's take a case of people shifting to fans from HVAC and for better energy efficiency shifting to LED bulbs from incandescent bulbs.

## 1.14 Demand-Side Management (DSM)

The concept of DSM is to able to manage power consumption at the consumers' side to meet their current and future requirements. These concepts could be promoted via a utility-sponsored program or through market segregator intermediaries such as ESCOs. Initially, DSM aims to create awareness programs such as customer or vendor rebates for efficient equipment and the like, and then later include time-of-use tariffs, interruptible tariffs, direct load control, and so on. Instead of building new power plants, promoting the use of energy reduction during peak demand periods can augment electricity capacity at a fraction of the cost. The cost of saved energy may be 10% of the added capacity for some systems. In addition to avoiding capacity costs and promoting energy efficiency at the customers' end, a utility can be brought closer to clients resulting in better customer service and a more efficient planning process. Demand-side management helps industries to be more energy efficient and competitive in this digital era.

### *1.14.1 Demand-Side Management in India*

The reasons for the problems of the Indian power sector are (i) high T & D losses, (ii) commercial losses due to energy theft, and (iii) low-efficiency equipment in the agricultural sector. Now, due to the extensive agreement, the restoration of finances in the power sector can be started by the demand side. A power sector company may be under DSM due to (a) outstripping demands to be provided by supply, (b) improving the revenues of the distribution companies, (c) improving the quality of power, and (d) mitigating the rise in tariffs.

### 1.14.2 Demand-Side Management and Power Quality

The link between power quality and its reliability is pretty self-evident. The usage of different power supplies in the system will reduce its efficiency. Voltage stabilizers, battery-run inverters, and irrigation pump sets employed by power consumers at low efficiency lead to coping strategies. Quality improvements have several positive implications and evidence shows that consumers are ready to pay higher prices if power quality and reliability are improved.

### 1.14.3 DSM in the Agricultural Sector

Irrigation system efficiency is about 20–50% in India. Therefore, the agricultural sector is highly inefficient with high water wastage. Over 27% of the total power consumption is used by the agricultural sector whereas revenue wise, the sector amounts to only 5–10% of all revenues because of the flat-rate pricing system for agricultural consumers.

Hence, the low tariff and old inefficient technology are the reasons for this water loss, and unsustainable exploitation of groundwater reserves and the high energy losses in irrigation.

## References

1. Ramachandran V, Shankar V, Shevgaonkar RK Transmission lines. Tata McGraw Hill, New Delhi
2. Modern trends and practices in power sub-transmission and distribution systems, vol-I. Central Board of Irrigation and Power, New Delhi
3. Modern trends and practices in power sub-transmission and distribution systems. vol-II. Central Board of Irrigation and Power, New Delhi
4. Gonen T (1988) Electric power transmission system engineering-analysis and design. A Wiley-Interscience Publication, Wiley
5. Gonen T (1986) Electric power distribution engineering. A Wiley-Interscience Publication, Wiley
6. Fischi R, Ountel WR (1972) Efficient method for computing electric power transmission network services. IEE Power Engg Soc Winter Power Mee Pap. C72 167
7. Electric Power Research Institute (1968) EHV transmission line reference book. EEI, New York
8. Electric Power Research Institute (1978) EHV transmission line reference book: 115–138 kV compact line design. EPRI Palo Alto, California
9. Electric Power Research Institute (1979) EHV transmission line reference book: HVDC to 600kV. EPRI Palo Alto, California
10. Uhalmann E (1975) Power transmission by DC. Springer-Verlag, Berlin

11. Deweese FC (1945) Transmission lines: design, construction and performance. McGraw Hill, New York
12. Rural Electrification Administration (1972) Transmission line manual. REA Bull. No. 62-1, U.S. Dept. of Agriculture, U.S. Govt. Printing Office, Washington, D.C.
13. Dwight HB (1954) Electrical elements of power transmission lines. MacMillan, New York
14. Edison Electric Institute (1968) EHV transmission line reference book. EEI, New York
15. Anderson PM (1973) Analysis of faulted power systems. Iowa State Univ. Press, Ames
16. Weedy BM (1980) Underground transmission of electric power. Wiley
17. Ash DO et al (1979) Conductor systems for overhead lines. Proc I.E.E. 126:333–339
18. Kimbark EW (1972) DC transmission, vol 1. Wiley, New York

# Chapter 2
# Distribution Planning and Optimization

## 2.1 System Planning Studies

Distribution planning is an important study for expansion of the power system network under load growth at the least cost. The main aim is to improve system performance, utilizing power in an optimal way and reducing the system operating costs, and also provide information to ensure an ample power supply and uninterrupted service [1]. Distribution planning must now include reliability/cost tradeoffs explicitly [2].

It is to be conducted periodically and it also provides the important characteristics of the power system, such as voltage and current profiles, losses, peak demand, annual usage, and load factor. It also helps the system engineer to manage system expansion and improve the performance of the system. In addition to the above the following studies are required for power system planning.

(a) Power precondition study
(b) Futuristic system planning
(c) Temporary system planning
(d) Coordination study
(e) Optimal conductor design
(f) Power factor improvements.

### 2.1.1 *Power Precondition Study*

Historical data such as total peak demand, yearly demand, consumption of yearly energy and procurement of energy include a minimum of the last five years' details collected in order to perform a power requirement study to forecast load growth. Using these data by plotting the graph gives a clear indication of upcoming



J. Raja et al., *Practices in Power System Management in India*,
Power Systems, https://doi.org/10.1007/978-981-10-2972-1_2

demand; it also shows current, past, and future annual losses and load factors yearly and monthly [3, 4].

### 2.1.2 Futuristic System Planning

Futuristic system planning is essential and guides the utility in an economical and orderly expansion of the distribution network also ensuring an adequate supply at the lowest cost to the consumer. It also provides information on system growth and helps utility management by identifying economically sound solutions in terms of building cost, existing facilities, system losses, and prevention of standard investments [5]. Understanding the map will help the system engineer and not affect the development of the general plant in making certain modifications in transmission lines or power substations. This study will provide a number of valuable benefits:

- Help the system engineer develop reliable and economical solutions.
- Increase system reliability and reduce the cost effectively.
- Determine the location and size of new substations and tie lines.
- Consider planned future investment for distribution automation such as SCADA, AMR, OMS, and GIS.

The following are required for detailed planning.

- Latest distribution map
- Major load locations
- Voltage and current investigation results
- Data relevant to present and forthcoming loads
- Power necessities study
- Summary of latest outage with details of indices (SAIFI, SAIDI)
- Current transmission facilities
- Availability of power in the future

System modeling is then done which will find:

- The ability of the system to supply power to cover maximum area, and voltage drop should not exceed more 6%.
- The total cost to be invested for the construction of a substation with all facilities including protection devices and lateral feeder circuits; the charge is based per square mile.
- Total losses are calculated based on planning load and associated cost for losses.
- The prospective level of system reliability indices (SAIFI, SAIDI).
- The total expenditure for the plan, including substation feeds, distribution feeder, and cost involved for system total losses.
- Cost-benefit analysis of the network reconfiguration.

The reliability level of the network can be planned as per the design. Depending upon the importance and location of the substation, line, and transformer, planning can be done either at the (N − 1), (N − 2), or (N − 3) level [6]. Typically for any distribution level network (N – 3) planning may not be required. Table 2.1 indicates the reliability level of a network that can be planned. Generally planners take into consideration the four nines (99.99) level network which means each consumer connected to the network may face on average 53 min of interruption during a year. Systems having substation automation with DMS and SCADA may reach a design level of five nines (99.999) which means an average of 5.3 min/year interruption to each consumer connected to the network.

There are a number of guidelines used for measurement, but the three most common and easiest measurements are SAIFI, SAIDI, and CAIDI as defined in IEEE Standard 1366.

(a) **System Average Interruption Index (SAIDI)**

This provides information about the average time customers face interruption and is normally calculated as the total number of customer interruption durations divided by the total number of customers served in the affected area; the unit for SAIDI is minutes or hours.

$$\text{SAIDI} = \frac{\sum \text{Customer Interruption Durations}}{\text{Total Number of Customer Served}} \tag{2.1}$$

(b) **System Average Interruption Frequency Index (SAIFI)**

The average of the occurrence of sustained interruptions in power per customer over a predefined area is called SAIFI. It is defined as the ratio between the sum of the total number of customers interrupted divided by the total number of customers served; the unit for SAIFI is a simple number.

$$\text{SAIFI} = \frac{\sum \text{Total Number of Customer Interruption}}{\text{Total Number of Customer Served}} \tag{2.2}$$

**Table 2.1** Availability reliability index

| Availability (%) reliability index | Nines | Annual interruption time |
|---|---|---|
| 90 | 1 | 36.5 days |
| 99 | 2 | 3.7 days |
| 99.9 | 3 | 8.8 h |
| 99.99 | 4 | 52.6 min |
| 99.999 | 5 | 5.3 min |
| 99.9999 | 6 | 31.5 s |
| 99.99999 | 7 | 3.2 s |
| 99.999999 | 8 | 0.3 s |
| 99.9999999 | 9 | 1.9 cycles |

(c) **Consumer Average Interruption Duration Index (CAIDI)**

This is the average time needed to reinstate power services to the average customer per sustained interruption. It is the sum of customer interruption durations divided by the sum of the total numbers of interrupted customers.

$$\text{CAIDI} = \frac{\text{SAIDI}}{\text{SAIFI}} \tag{2.3}$$

## *2.1.3 Temporary System Planning*

Temporary system planning plays a vital role in completing the systematic growth of the system and also fulfilling the requirement of load growth for the entire distribution system [7].

(a) **Voltage Drop Study**
Temporary systems study is achieved by the performance of the voltage drop study. This study helps to uncover strong and weak points of load voltages at different load points.
(b) **Voltage Drop Factors**
The study discovers voltage drops at various load points with different conductor sizes.

## *2.1.4 Coordination Study*

The coordination study analyzes all or part of the distribution system commonly called a sectionalizing study. The purpose of this study is to find the ability to place a sectionalizing device and its selection choice is on the basis of fault current. The combination of the coordination study along with the voltage drop and reliability study is used to integrate major system changes in load or system configuration.

This main objectives of this study are as follows.

- Help to reduce the per customer power outage per year and also indirectly reduce the service restoration cost after the outage.
- Reduce the frequent failure of apparatus and in turn increase the lifetime of the power equipment.
- Reduce the probability of unsafe voltage on objects grounded with the neutral system.

The above objectives are achieved by the use of proper selection and location of sectionalizing devices on the distribution system. By the proper selection of sectionalizing devices, the hazardously adverse line operating characteristics can be minimized to an acceptable range but the outages, equipment damage restoration expenditure, and unsafe voltage can never be completely eliminated. In terms of paperwork and measurements, in modern power systems computational techniques using digital computers for the calculation of load and fault currents and thus the performance of the coordination study have been radically simplified. A coordination study is performed through the following four steps.

(a) Accrue and process data.
(b) Test sectionalizing for coordination and correct ratings.
(c) Alter sectionalizing to correct inadequacies that would correct the reliability of the network to desired level.
(d) Document findings as required.

Steps (a) and (b) are executed with the help of digital computers and (b) and (c) are repeated as and when required. These actions are periodic tasks that do not require any engineering judgment. Procedure (c) could be executed using digital computers with prompt software; it would be hard to duplicate sectionalizing situations. Final documentation calls for the engineer's communication skills. The necessary data essential to complete a coordination study are:

(a) Correct system model with types of device and location of each sectionalizing point.
(b) Sectionalize point peak load current.
(c) Each point maximum/minimum fault current.
(d) All devices with their rating and time current characteristic curves.

System mapping shows the system configuration, consumer location, and existing sectionalizing points. For the calculations of load and fault current, this model must display the data's information including line loading in a particular section, with its distance and phasing. It is suggested that the line and its fault current values be added to the sectionalizing study map to assist sectionalizing point evaluations.

The final document should include a system map with locations of sectionalizing points for communication and it is mandated to monitor and verify the computer results. The customized system circuit diagram will be sufficient for computer-aided study to incorporate sectionalizing device locations. For this study it is necessary to use all the system voltage, continuous and interrupting current, minimum pickup ratings, and time current characteristics. These data are constant and permanently stored on computer, because the data are useful and necessary for all studies, except for addition of devices. When a proper coordination study is implemented in association with a structure work plan it will save time and expense significantly. Ample data are necessary for a coordination study to perform a construction work plan. To decide if system sectionalizing is sufficient, the device at each sectionalizing point should be tested for accurate voltage, interrupting current, continuous current, pickup current ratings, and also synchronization with adjacent devices.

### 2.1.5 *Optimal Conductor Design*

The purpose of an optimal conductor design is to reduce energy losses to the lowest possible level. While designing the primary lines for an electrical distribution system, electrically conductive adhesive (ECA) minimize the cost, and more important, the specified conductor size should be adequate from a voltage drop standpoint. A shortened technique of economic comparison can be developed using data and methods that result in a graphic depiction of the total cost to own and operate a variety of lines at various load levels [8]. The following data are necessary to carry out the economic analysis.

(a) Three-phase conductor cost with various sizes from manufacturer
(b) Electrical system fixed costs including O&M expenditure, represented as a percentage of plant value
(c) System load factor
(d) Amount of energy required and demand for the system

Dissimilar conductors' cost of energy losses and their savings (in energy losses) are calculated at different load levels. Finally, a graph should be drawn in comparison indicative of the most economical conductor size required for each particular load level.

### 2.1.6 *Power Factor Improvement*

The power factor (PF) is defined as the ratio of real power to apparent power or the rate of useful working current to the total current in the line and is calculated by the formula:

$$\text{Power Factor} = \frac{\text{kW}}{\text{kVA}}$$

When the PF is below 90%, a penalty is imposed by the supplier/DISCOM to the consumer because under these circumstances loss in energy is too high. Capacitors are installed in online strategic locations for correcting the PF. The installed capacitor establishes power factor improvement, energy loss reduction, system voltage rise, service reliability, and prevention of expensive line conversion cost.

## 2.2 Central Electricity Authority of India (CEA) Guidelines on Distribution Network Planning

Insufficient network planning is one of the causes for disorganized scientific development of the distribution system. The DISCOMs should plan and expand the network according to demand forecasting on a medium- and long-term basis and for determining the need for system extension and enhancement to meet the load growth. The utility should prepare a prospective network plan for a 10-year period. This should become part of the conditional criteria for sanction of grants under various programs.

## 2.3 Operations Overview

The functions of a distribution network are to transform and distribute electricity throughout the DISCOMs network. This system should include necessary electrical equipment to obtain the required power and convert the different voltage levels and to manage and distribute the power to the various electrical load levels. Attention must be focused on O&M and planning of the distribution system for the various short- and long-term needs.

### 2.3.1 *Operations Management*

Electrical distribution is a unique system, thus each DISCOM has separate adequate operating procedures that are particular to a given site and the system of concern.

### 2.3.2 *Operating Procedures (Preparation of Operation Manual)*

A separate manual is provided for the operating procedure of each electrical distribution system component. It should contain instructions for sequence switching of the main and feeder switching devices. Often operating manuals labeling the details of various portions of the system for complex systems are required. The manual should contain up-to-date information about:

- Equipment details used in the network.
- Reflection of the operating modes used.
- Explain sequence procedure to energize the whole system from the incoming point of connection to the lowest utilization voltage level.
- Must have drawings, figures, tables, and equipment drawings so that additional documents should not be needed for operation.
- Contain information about the distribution system functions.
- Describe the electrical distribution functions and operation capability/limitations of various components involved in the electrical distribution system.
- Include emergency ratings of all the distribution components involved in the particular system such as transformers, distribution panel, motor control center, feeder cable, and switchgear.

- Contain the information relating to distribution system operating modes under a collection of system configurations and describe the sequential procedure necessary to energize the whole system and sequential procedure necessary to alter the configuration of the system.
- Explain the purpose of various alarm, trouble, fault detection, and equipment shutdown systems provided.
- Describe the analysis of various annunciator, alarm, protective relay, and equipment shutdown systems, together with various checklists, as necessary, to point out likely causes of the trouble indication.
- Indication with effect on system operation in trouble and explain the ways of service restoration of the fault-affected area and contain an optional course of action.
- Contain latest single line diagram (SLD) of the whole distribution network with each equipment size and ratings and indicate the normal operating position of circuit switching devices that may be used to modify system operating configurations showing positions of all bus tie breakers and switches, circuit sectionalizing devices, and incoming breakers from multiple sources if all sources are not energized concurrently.
- When equipment is replaced or modified, the system drawing must be revised indicating changes in the drawing made accordingly. Subsequently the operating manual should be revised supposing the alteration may affect the system operation modes.

### *2.3.3 Routine Operation*

With the introduction of automated devices in the network a system can be operated automatically; moreover, well-designed distribution equipment does not require frequent operating changes [9]. Data collection of major parameters such as voltage, current, energy, and power factors at pertinent points of the system can be helpful in the repetitive operation of the system as well as future planning exercises. In modern days load flow studies are analyzed with the help of faster rate inexpensive computers, which is very useful in determining the optimum operating configurations of the distribution system.

## 2.4 System Disturbances or Outage

Frequent changes in the distribution system and occurrence of fault or overload that result in outage or disturbance can be surge, impulse, noise, and phase shift and sag, among others. Based on previous studies, only one phase of a three-phase system subjected to frequent fault and the occurrence rate is around 80–85%; almost 90% of the disturbances are less than one second. Disturbances are categorized based on their time duration; when it is over one minute it is usually classified as an outage.

### 2.4.1 Disturbance Categorizations

- **One Second to One Minute**: The effect of these disturbances is 50–100% of voltage loss on a single phase or more than one phase in the distribution system. These are categorized as severe faults.
- **10 to 40 Cycles**: These faults are grouped as sag or surge. Due to the slow speed of switching devices, operation delay in tap changers on the transformer, voltage regulator, and starting of motor delay, these disturbances occur in the distribution system.
- **0 to 8 Cycles**: These disturbances are classified as surge or sag, because of high-speed switching devices and sensitive fuses. In a three-phase, a single-phase load may cause a surge on the unloaded phases or remaining phases while causing sag on the loaded phase.
- **0.001 to 1 Cycles**: These disturbances are typically stimulated in the distribution system due to capacitor switching, short duration fault, and surge arrester operation.
- **Less than 0.001 Cycle**: These are most severe disturbances caused by lightning, switching of nearby loads, and electrostatic discharge, and normally known as impulse.

The utility supplying power to the customer rarely obtains power quality and continuity; the effect of fluctuation in supply can be reduced to a tolerable level by adopting the following modifications in (i) design of utilization equipment, (ii) ensuring the distribution system is well suited to the utilization equipment, and (iii) both system and equipment standard should be accurate with each other. A buffer circuit should be created between the prime source and utilization equipment to interrupt the continuous supply system. The buffer circuit acts as an external source of disturbances, but could raise the magnitude of load-induced disturbances.

Filters, power conditioners, surge arresters, capacitors, solid-state motor starters, adjustable frequency motor drives, uninterruptible power supply systems, isolating transformers, and revisions to grounding systems are the most commonly used solutions to suppress electrical disturbances. Disturbances are classified by their time duration (not less than a second), and according to the disturbance the systems are designed and have installed proper protective devices so that these devices operate automatically without operator intervention to identify and clear faults.

An automatic alarm should be integrated with protection devices to indicate the occurrence of a fault to the operation or maintenance personnel. This will help maintenance personnel restore electrical devices in an affected area more quickly. The maintenance personnel should follow the systematic procedure and maintain the records as follows.

(a) Maintain the document, listing the problem and reason for outage.
(b) Modify configuration of the system and restore the service, as per the requirement.
(c) Prepare the timetable for repairs, if necessary, and arrange suitable repairs or equipment replacements.
(d) Make sure the system is in normal operational configuration after the restoration.

The above procedure should be completed in an appropriate and safe method. After the occurrence of a fault a detailed investigation should be made to find the cause of failure; accordingly the operator should remove the faulty equipment from service. After the complete system safety procedure is followed, power is once again restored, including inspection of the suspected fault location, inspection of possible damage to equipment, and ensuring OFF position of switch devices, so that the unanticipated reversal of power from some of the equipment will not cause harm to the workforce or apparatus. All of these steps must be part of the operating procedures manual.

## 2.5 Power Quality

This is defined as the grid's ability to supply stable and reliable power. In other words, the grid should maintain a standard frequency and voltage or within the tolerance [10].

- **Voltage Regulation**: It can be improved by use of off-load as well as on-load tap changers, installations of capacitors, and PF correcting devices.
- **Harmonics**: With the help of the latest technology in solid-state switching devices implemented in distribution equipment, harmonics problems are reduced significantly and maintain the total harmonic distortion (THD) within prescribed limits, and other quality parameters are properly maintained with the help of variable frequency drives, rectifier power supplies, high- frequency power supplies, uninterruptible power systems, and arc discharge lamps.
- **Frequency**: The difference in generation and demand on the system causes frequency variation in the network, and it must maintain the standard value as closely as possible.
- **Voltage Imbalance**: Uneven distribution of single-phase loads on each phase will cause voltage imbalance in the system; it is normally limited to a maximum of 3%. For balancing the voltage, essential efforts should adopt the single-phase loads that should evenly distribute all three phases [11].

## 2.6 Maintenance Planning and Implementation

Maintenance is a regular action in response to disturbances, failures, and other unfortunate events and is normally carried out by maintenance personnel. It is a kind of method that may be severe and even failure of a relatively small component may affect the overall capacity when a plant is under operation or when there is construction of a new line and power plant assembly [12]. Preventive maintenance action ensures operation continuity and lowers the risk of accidental outages. Preplanned shutdown for maintenance takes place during the period when the system is in least usage or inactive. As a result, earlier stage detection of problems and accordingly proper action can be taken before any damage occurs. Table 2.2 shows the number of failures against maintenance quality for all equipment classes as per the IEEE Industrial Commercial Power Systems Committee.

The effectiveness of the maintenance survey indicates almost 33% of all maintenance cost is wasted for improper maintenance. The main reason is lack of actual data to quantify the actual need for maintenance of plant machinery, equipment, and systems.

### 2.6.1 *Evolving World-Class Maintenance*

The path to maturity is developing all over the world in the stages shown in Fig. 2.1.

The main aim of an electric utility is to improve system reliability; to reduce O&M costs are the top priorities. Electrical maintenance varies from time to time as a result of competition and costs, and equipment difficulties are increasing as well as regulatory approval for O&M. Figure 2.2 shows maintenance themes; in general maintenance is either scheduled or unexpected.

Corrective maintenance is an unexpected approach and is carried out after the fault occurrence. The aim is to reinstate an item such that it can carry out the regular required function in the system. Depending upon the nature of maintenance, it can be categorized into two groups: preventive or scheduled maintenance or regular maintenance. Under this category there are four basic tasks:

**Table 2.2** Number of failures versus maintenance quality

| S. No. | Maintenance quality | All causes | Insufficient maintenance | Failure due to inadequate maintenance (%) |
|---|---|---|---|---|
| 1. | Excellent | 311 | 36 | 11.6 |
| 2. | Fair | 853 | 154 | 18.1 |
| 3. | Poor | 67 | 22 | 32.8 |
| | Total | 1231 | 212 | 17.2 |

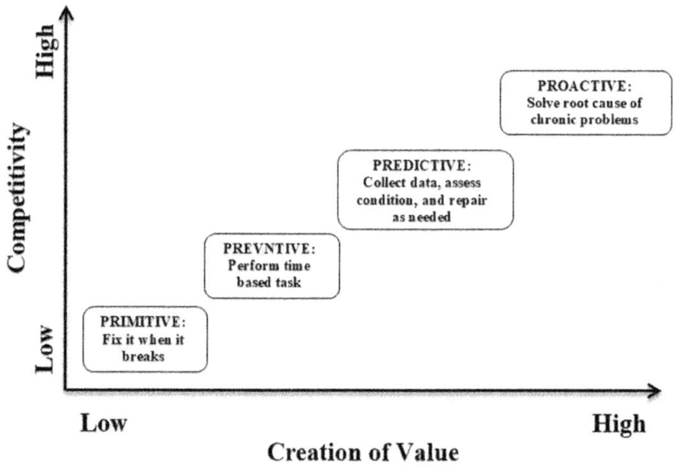


**Fig. 2.1** World-class maintenance

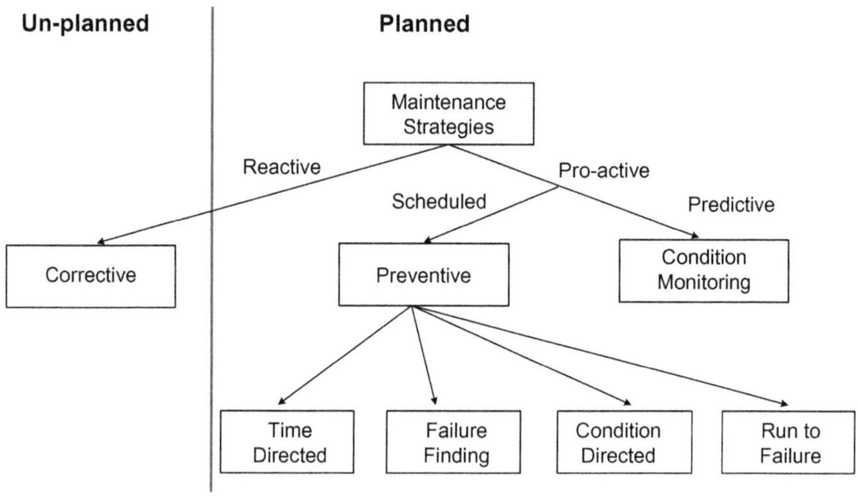


**Fig. 2.2** Maintenance strategies

1. **Time Directed**: Involves number of operations, operating hours, or seasonal change
2. **Failure Finding**: Used to protect the equipment; able to find equipment failure
3. **Condition Directed**: Used to prevent incipient failure from becoming a real failure when sustained acceptable operation cannot be ensured
4. **Run to Failure**: Optional task that can be selected only in the event a technically correct and cost-effective task cannot be identified.

### 2.6.2 *Main Causes of Electrical Failure*

An effective maintenance aims to minimize the failures in the electrical system. The accumulation of dust, dirt, or moisture can be reduced by keeping the apparatus in clean and dry condition; similarly, loose connections and friction of moving parts can be minimized by keeping contacts stable and supplying proper lubricant oil in moving parts particularly where friction occurs.

**Dust and Dirt Accumulation**
Degradation of insulation and flashover may happen in the system due to accumulation of chemical dust, lint, and the deposit of oil mist; the particles become conductive when combined with moisture on insulation. Accumulation of dirt builds on transformer coils, and relay lines will reduce the air flow, in turn increasing the operating temperature leading to equipment failure or reducing the lifetime of the equipment. Similarly, dust accumulation in the case of outdoor insulators of overhead lines and substation equipment can meet flashovers. Contamination cannot be avoided in coastal areas and some industrial areas but with the help of periodically scheduled equipment cleaning these contaminations are definitely reduced.

**Presence of Moisture**
Condensed moisture can lead to oxidation, degradation of insulation, and failure in electrical equipment. Free condensation produced by high humidity may lead to short-circuits and failure of the electrical equipment. Rarely does electrical equipment operate in the dry atmospheric condition which is, however, the ideal case. To minimize moisture, proper enclosure and space heaters can be used.

**Loose Connections**
Connections should be tight and torque to the manufacturer's suggested values. Joint failure usually happens due to creep/cold flow by load cycles. The most common areas of loose connections are cable connections, fuse clips with circuit breakers, and contactors; all these areas must be regularly checked.

**Friction**
Friction can affect the free movement of electrical devices; for example, in circuit breaker (CB) friction can reduce the operating speed, and it is of prime importance in circuit breakers. Furthermore, dirt accumulation on parts can cause scratches and arcing or burning may occur. All devices should be properly lubricated with specific type and grades, unless proscribed by the manufacturer. Insulation is attacked by dust and other contaminants on oil and grease. Checking the mechanical operation of devices and manually or electrically operating any device that seldom operates should be standard practice.

### 2.6.3 Breakdown Maintenance or Repair

It is also termed the "run to failure" maintenance technique. Even today, many utilities follow the "run it until it breaks" philosophy. All too often we hear statements such as, "We can't take that out of service," or "We've never had a problem so why bother to perform maintenance." The cost involved for failure includes outage and its repair cost is definitely less than the preventive maintenance cost. Surprisingly, breakdown maintenance employs minimal preventive maintenance techniques, however, action is taken only when urgently and immediately required. The following shows some common data items and how they are put to use in a breakdown maintenance approach.

- **Load Current**: Transformer or other equipment replaced at first outage
- **Voltage**: Tap changers adjusted, if provided
- **Temperature**: Symptomatic treatment such as checking the cooling system or adding fans

Breakdown maintenance may not involve any datapoint analysis; it will be cost effective until no disastrous failures occur. Such an approach leaves the system open to major disasters because no precautions are taken to avert them. Perilous conditions may exist with no way to predict them.

### 2.6.4 Preventive Maintenance

The most commonly accepted approach to maintain electrical equipment is no doubt preventive maintenance. Time-based complete routine tests are applied to offline equipment. The insulation resistance (IR) test, protective device calibration checks, power factor test, and time travel analysis for circuit breakers are used to evaluate present system conditions.

With an example, the following briefly explains how preventive maintenance is used.

- **Insulation resistance**: Resistance values are calculated and checked as per industry norms. Doubtful insulation is scheduled for retest and bad insulation is replaced immediately.
- **Insulation power factor**: Treated in a similar manner as above.
- **Protective devices settings**: Equipment is also cleaned and adjusted mechanically to meet calibration settings provided by engineers.
- **CB time/travel analysis**: Breakers are adjusted to manufacturer's specifications.

The two major changes between preventive and breakdown maintenance programs in terms of the way the collected data are used are:

- Data are usually collected during both online and offline times. Offline times are intentionally scheduled for the implementation of preventive maintenance procedures.
- The equipment that requires repair is scheduled for outages and to implement the said repairs.

The principal problems with preventive maintenance programs can only be statistically evaluated and cannot be economically evaluated. The cost expenditure is directly compared between maintenance and expenditure for outage. However, the cost of expenditure cannot be calculated except when an unplanned outage occurs.

### 2.6.5 *Predictive Maintenance (PDM)*

As proved, PDM is better and cost effective. For example, regarding the dissolved gas analysis in oil, practical experience has set empirical bounds for the rate of change and amount of combustible gases in the insulating oil. Based on extensive analysis, we know that if the amount of acetylene rises above a predefined standard concentration or if the rate of production of said gas increases beyond a certain value then arcing will undoubtedly occur in a transformer. Thus we can safely predict the occurrence of arcing if the analysis shows an exceeded amount of acetylene is found. On further analysis, prediction of the transformer life can be calculated. An emergency outage can be scheduled for the faulty transformer; then the cause can be isolated and repaired. PDM is not a general systemwide strategy but an equipment-specific type of approach; that is, most companies and agencies do not use predictive techniques throughout the system. To perform predictive procedures, these companies use equipment or specialty vendors on their critical transformers. Infrared scans are performed on whole systems. The common types of PDM procedures employed in current power systems are summarized below.

- Dissolved Gas Content: A comparison of industry standards and gas content and percentage increase are computed. If equipment exceeds these standards it is scheduled for offline repair.
- Thermographic Scan Temperatures: Temperature rise scans are associated with industry norms.

Suppose the equipment exceeds offline repair scheduled norms. It shows the maintenance engineer how such programs work. The procedure for a PDM program is summarized below:

- Important parameters are measured and analyzed.
- It gives a clear idea about general standards of a particular component with respect to potential problems associated with the equipment.
- Equipment that fails the standards is strategically planned for repair or retesting.

PDM methods are commonly part of a preventive maintenance program itself. Hence, no sudden outage occurs, repairs are scheduled, and major failures are dodged; that is why this is one of the best cost-attractive maintenance programs.

### 2.6.6 Condition-Based Maintenance (CBM)

Condition-based maintenance (CBM) is almost similar to PDM. In CBM, the collected data through certain intervals are compared to statistical norms, both arithmetic mean and logical trends; it uses both online and offline test data, thus it is more comprehensive than PDM. The main advantage of CBM is the manner of data analysis [13]. Table 2.3 illustrates these principles. Future maintenance procedures on the equipment are determined by the results obtained from data.

The first column of Table 2.3 lists the most important differences between CBM and other types of maintenance. The next schedules the maintenance of the equipment cycle that may be skipped, provided the equipment is in strangely good condition as directed by the test. Then the overall maintenance effort is reduced significantly, because only a portion of the system equipment will show problems in the test results at any one interval, possibly as little as 20% of pre-CBM levels. Such reductions are only possible because the equipment which is shown, by test data, to need additional servicing would be watched for in the next maintenance cycle. Equipment that has been tested once in two years may now need attention and only need a check every five to ten years.

### 2.6.7 Reliability-Centered Maintenance (RCM)

RCM is an algorithm that helps the DISCOMs to regulate the optimal preventive maintenance tasks necessary to report critical equipment failures without compromising service reliability. It is the combination of corrective, predictive, and preventive maintenance, and utilized on these approaches where suitable, based on the

**Table 2.3** Condition-based maintenance

| Results within standards | Reasonable deviation | Severe deviation |
|---|---|---|
| Short-term service checks | Perform more in-depth procedures during this interval | Scheduling of interim tests before the next routine maintenance cycle |
| Completely skip the next scheduled maintenance | Include as usual in next maintenance interval | Removal of online equipment for disassembly and rebuilding |

significance and frequency of equipment failures. Therefore, optimization of the maintenance program is economical and reliable. It mainly aims to prevent most severe faults in the system. PDM and RCM analysis balance each other; their performances are parallel for optimization of the power system. It ranks the operation of critically based equipment, the interruption cost in revenue and customer loss, safety, cost of repair, and so on. Data measurements used in PDM and RCM are the same but require more training to save extra maintenance resources by spending less effort on lesser critical machinery.

### *2.6.8 Premonsoon Maintenance*

Because most of the distribution network is overhead/overground in nature, it is susceptible to various faults due to storms, lightning strikes, heavy rains, and floods. Indoor and underground cables are also subjected to tough times but to a comparatively lesser extent. It is therefore a general practice followed by most distribution utilities to plan and carry out thorough maintenance of the distribution network components of lines and substations. A snapshot of maintenance carried on different equipment is as below.

**Distribution transformer**
Check oil level and leakages and rectify.
Check silica gel breather and replenish with blue colored gel.
Check and tighten all current-carrying clamps and connectors (tighten or replace if need be).
Measure dielectric strength of the transformer oil and carry out filtering if need be.
Check all earthing connections for tightness and continuity (measure and bring value below 5 Ω).
Check LA connections and tighten.
Check working of primary and secondary side protections and rectify if need be.
Provide fresh crimping to low tension (LT) cable connections to bushings.

**Overhead lines**
Straighten the bent poles and replace cracked PSC poles.
Tighten stays.
Check and provide ant-climbing devices and caution notices.
Check insulators for cracks.
Redo the binding of pin insulators (if damaged).
Check and rectify earthing of poles and all metallic fittings on it.
Redo muffing if damaged.
Check jump ring clamps, connectors for health.
Check multiple neutral earthing for LT lines and neutral continuity.
Ensure tree-cutting is done as per norms (vegetation management).

**Substation equipment (33/11 or 66/11 kV)**

Maintenance of the equipment in the substations is to be carried out routinely as per schedule inasmuch as breakdown on such equipment may result in overly long duration interruptions affecting a bigger geographical area and/or more consumers. Before a monsoon it is necessary to ensure that the maintenance has been carried out per the time schedule. The remarks in the history cards maintained for such equipment are to be attended to for any remedial action before the onset of monsoon.

## 2.7 New Technology in Maintenance

Today many technologies are being examined and used to control the electrical equipment's condition, helping in condition-based maintenance. A few tools and techniques are as follows.

### *2.7.1 Ultrasonic Noise Analysis Technique*

The prolonged sound in the equipment may be a sign of air, gas, or steam leaks and can also be a result of friction between moving parts.

### *2.7.2 Incomplete Discharge Detection*

Sensors used in the power system detect initial insulation breakdown in any equipment; they are also used to detect emerging failure so that in future no significant damage occurs.

### *2.7.3 Transformer Dissolved Gas Analysis*

This analysis is one of the methods used to detect the amount of dissolved gas content in transformer oil or any other any abnormalities. A gas-level indicator shows the maintenance engineer the state of the transformer essential for maintenance or potential failure.

### *2.7.4 Infrared Thermography*

Infrared cameras are used in thermography surveys to detect hotspots at various clamps and connectors in distribution substations and lines. They show the real-time measurement of the four key gases—carbon monoxide, hydrogen, acetylene, and ethylene—which are known for fault currents in the transformer.

### 2.7.5 *Hotline Maintenance*

This is the process of maintenance at transmission or distribution lines in live condition without necessitating any shutdown (Fig. 2.3). Sufficient insulation has been provided by the technology during maintenance. This is very high-risk technology to be executed only by properly trained and experienced personnel. Furthermore, an extra sensor is deployed to detect the moisture level in transmission lines; as a result it may reduce the dielectric strength and failures. These sensors are used to measure transformer loading, recording the development of gas and moisture content as a function of load. With the moisture sensor and load current monitor, an accurate criterion for transformer loading under stressful conditions can be developed rather than depending upon conservative ratings.

## 2.8 Integrated Approach

DISCOMs fail to integrate and process expediently a quick evaluation of the large amount of data collected, thus leading to delayed O&M recommendations, and thereby adding cost for data collection and still failing to provide cost benefit from effective maintenance. RCM and PDM analyses balance each other, and excellent optimization is achieved when they are performed simultaneously. A comprehensive approach for T&D systems ensuring a constant and reliable supply should include the substation and equipment. Please see Fig. 2.4.

### 2.8.1 *Case Study and International Practices for Equipment Maintenance*

In India, since the initiation of power sector reforms many vertically integrated utilities/state electricity boards (SEBs) have been unbundled. The role of newly

**Fig. 2.3** Hot line maintenance work

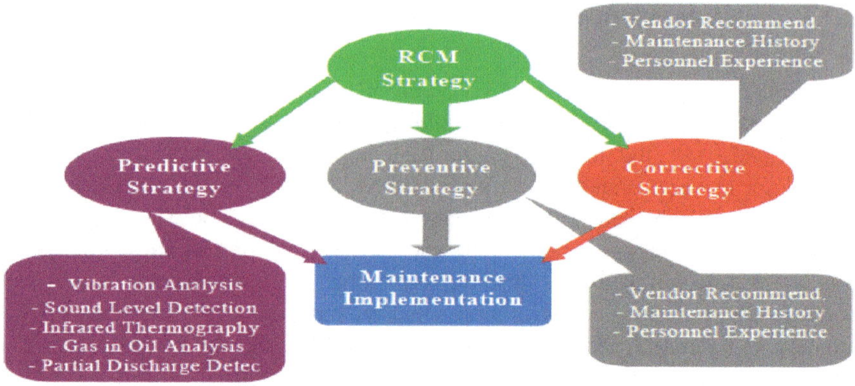


**Fig. 2.4** A representative figure displaying the integrated approach

formed distribution utilities has become more focused on meeting the needs of retail consumers. These companies have inherited very old and weak networks facing frequent breakdowns of lines, cables, and transformers. Most of the DISCOMs are faced with challenges of fulfilling growing consumer expectations in the area of providing a continuous and good quality power supply. This is to be achieved by upgrading and toning the distribution network and adopting best practices in the area of network augmentation plans and adapting to the best distribution and maintenance standard practices.

A power distribution company in its journey towards "Service Excellence to Consumers" needs to set a corporate target to be achieved by a predefined horizon year (may be in a span of four to six years) that relates to improving the service delivery mechanism to consumers. This necessitates formulation of a strategic reliability and maintenance roadmap.

The activity starts with performing an assessment of existing reliability and maintenance practices within the utility and identifying a desired future state based on benchmarking with best-in-class utility practices. The roadmap is prepared with an implementation schedule covering details of a year-on-year plan of activities, expenses, and achievements until the horizon year is reached.

### *2.8.2 Understanding "As-Is" Scenario of Distribution Network Assets*

At the beginning of the activity it is necessary to know the present health of the various network components of the detailed equipment audit to arrive at a decision of run as it is, repair and then run, or replace [14, 15]. Such a network audit is conducted through OEMs (original equipment manufacturers) and deploys teams of in-house engineers. This covers the details of 66 or 33 kV substation equipment such as power transformers, circuit breakers, isolators, capacitor banks, battery and chargers, and the like, as well as overhead and underground HV and LV feeders, distribution transformer substations with control gears, service cables, and meters.

### 2.8.3 Formulation of the CAPEX Plan

Deciding on the goals to be achieved at the end of the horizon year includes in the area of AT&C loss reduction, reliability improvement, and catering to consumer growth by adequate interventions of IT enabling systems and distribution automation with SCADA. Having frozen the yearly activity, expenses, and achievement phasing program the utility can now go ahead with formulation of reliability and maintenance roadmaps based on international best practices.

### 2.8.4 Typical Causes of Poor Reliability

- Unavailability of power as required
- Delay in postfault rectification
- Equipment failure
- Trees coming in contact with overhead lines
- Lightning
- Animals (reptiles and birds)
- Overloads
- Traffic accidents
- Digging instances

### 2.8.5 Basic Ways to Improve Reliability

Prevent failure from occurring.

- Tree-trimming program
- Use of ABC (aerial bunched cables)
- Infrared feeder inspection program
- Conductor/cable replacement program
- Increased lightning protection
- Transformer load management programs

Reduce the number of affected consumers.

- Fuses on laterals
- Reclosures on main trunk

- Automatic sectionalizers on taps
- System reconfiguration

After fault occurs restore supply to more consumers by quick alternate switching.

- Use of normally closed and open switches on the feeders.
- Adhere to feeder and substation automation.

Locate fault quickly.

Install OMS (Outage Management System) which can have the following features.

Real-time circuit topology, automated trouble call processing, automatic interruption and outage data capture, integrated crew dispatch, fault passage indicators, links to SCADA, infer fault location based on fault magnitude, prefault current and feeder circuit model, greater accuracy using clearing times TCC (time current characteristics) curves.

Repair fault more quickly.

Mobile crew arrangement.

Performance management.

Summary:

- Reliability can be substantially improved at low cost.
- Use of rigor and discipline is required.
- Use an integrated approach.
- Need not focus obsessively on reliability indices.
- Start collecting good data.

Traditionally a utility mindset focuses on keeping the cost within approved regulatory provisions and maximizing performance without taking any risks. This can be attributed to the ageing infrastructure, increased regulatory oversight, and pressure to improve reliability, reduce costs, and increase earnings. There is a need to shift from this mindset to a new approach that envisages minimizing cost to achieve prudent performance by managing risks. The motives that drive this would be:

- Identification of efficiency gains
- Improvement of asset replacement technique
- Life extension program for assets
- Reducing negative surprises.

## 2.9 Reliability Data

Many utilities still perform their reliability analysis based on manually prepared paper reports as part of the data collection of planned outages, interruptions, breakdowns, and load shedding. To avoid the complexity of data collection, today all DISCOMs have installed an outage management system (OMS) that tracks the data and consumer interruptions as they occur. Thus knowledge of reliability indices such as SAIDI and SAIFI is more accurate. This requires GIS mapping of all network assets until consumer indexing. With the R-APDRP part A and B projects funding many utilities can avail themselves of this facility. Effective reliability management requires good asset data. This can be maintained through installation of ERP (enterprise resource planning) software. Otherwise such detailed inspection, maintenance, and failure history for equipment is stored in a computerized maintenance management system (CMMS).

### 2.9.1 Utility Trend in Reliability

- Increasing consumer sensitivity where consumers' expectations on reduction in interruptions are increasing
- Tougher regulatory targets of reliability prescribed while deciding tariffs on ARR submissions
- Increasing use of SAIFI and SAIDI in deciding employee incentives
- Increasing pressure by regulators to improve worst performing feeders
- Increased role of OMS so there is no chance of data manipulation
- Increasing pressure for reducing O&M cost
- Increasing focus for changeover from time-based maintenance towards condition-based and reliability-centered maintenance
- Increasing use of auto-reclosures and line sectionalizers on feeders
- Increasing use of automation
- Increasing use of cost-benefit analysis
- Increasing tendency to take more risks

Based on these trends and considering the activities planned in the CAPEX plan a typical roadmap for reliability improvement from its present status to the desired state is prepared.

### 2.9.2 Reliability Roadmap

It should address the following major goals to reach the desired future state.

- Dramatically improve overall reliability as measured by SAIDI and SAIFI.
- Demonstrate to all stakeholders that the expenditure incurred is in tune with the benefits received and maintain the tariff at a reasonable level.
- Establish business systems that would be able to achieve global benchmarks.

## 2.10 International Benchmarking

This section contains a comparison of maintenance and inspection processes and frequencies for high-voltage substation equipment from a number of utilities worldwide. The utilities that are represented in this section represent best-in-class utilities in Europe, North America, and the Middle East. International best practices for utility T&D substations for a variety of equipment types are contained in Tables 2.4, 2.5, 2.6, 2.7, 2.8, 2.9, 2.10, 2.11, 2.12, 2.13, 2.14, 2.15 and 2.16.

Differences in operating conditions, equipment condition, and the impact of failure of certain equipment or equipment types will always account for some differences. The challenge for the utility is to continue to maintain all its equipment and systems in a systematic manner and one that maximizes equipment reliability and availability. The utility has the goal to migrate from a time-based to a reliability-centered maintenance program.

**Table 2.4** International practices, minimum oil CB inspections, diagnostics, and maintenance

| S. No. | Inspections | Frequency |
|---|---|---|
| 1. | General visual inspection (looking for obvious defects) | Once per month |
| 2. | Visual check for leaks/contamination/moisture/loose connections | Once every six months |
| 3. | Visual check of terminations | Once every six months |
| 4. | Functional test | Annually |
| 5. | Infrared scan | Annually |
| S. No. | Diagnostics | Frequency |
| 1. | Measurement of contact resistance | Once every three years |
| 2. | Measurement of switching time and simultaneity of the phases. Lubricate moving parts at this time | Once every three years |
| 3. | Measure spring charging time for spring-operated breakers | Once every three years |
| 4. | Random check (−10%) of the condition of the oil per station (moisture, color, breakdown voltage) | Once every six years |
| 5. | Partial discharge measurements | Once every six years |
| S. No. | Maintenance | Frequency |
| 1. | Revision of breaker chamber: a. Replace contacts, b. Clean the chamber, c. Replace oil | After a short-circuit |
| 2. | Cleaning of bushings | Once every three to five years, also frequent cleaning is required for worst ambient conditions |

**Table 2.5** International practices, bulk oil CB inspections, diagnostics, and maintenance

| S. No. | Inspections | Frequency |
|---|---|---|
| 1. | General visual inspection (looking for obvious defects) | Once per month |
| 2. | Visual check for leaks/contamination/moisture/loose connections | Once every six months |
| 3. | Visual check of terminations | Once every six months |
| 4. | Functional test | Annually |
| S. No. | Diagnostics | Frequency |
| 1. | Measurement of contact resistance | Once every three years |
| 2. | Measurement of switching time and simultaneity of the phases Lubricate moving parts at this time | Once every three years |
| 3. | Measure spring charging time for spring-operated breakers | Once every three years |
| 4. | Random check (−10%) of the condition of the oil per station (moisture, color, breakdown voltage) | Once every six years |
| 5. | Partial discharge measurements | Once every six years |
| S. No. | Maintenance | Frequency |
| 1. | Visual check of contacts | After a short-circuit |
| 2. | Check of oil color | After a short-circuit |

**Table 2.6** International practices, $SF_6$ CB inspections, diagnostics, and maintenance

| S. No. | Inspections | Frequency |
|---|---|---|
| 1. | General visual inspection (looking for obvious defects) | Once per month |
| 2. | Visual check for leaks/contamination/moisture/ loose connections | Once every six months |
| 3. | Visual check of terminations | Once every six months |
| 4. | Functional test | Annually |
| 5. | Check to trace $SF_6$ leaks | Annually |
| 6. | Check foundations, grounds, and paint | Once every five years |
| S. No. | Diagnostics | Frequency |
| 1. | Measurement of contact resistance | Once even, three–five years |
| 2. | Measurement of switching time and simultaneity of the phases. Lubricate moving parts at this time | Once every three years |
| 3. | Measure spring charging time for spring-operated breakers | Once every three years |
| 4. | Partial discharge measurements on the cable terminations | Once every six years |
| 5. | Power factor insulation test | Once every five years |
| 6. | Moisture test on gas | Once every five years |

(continued)

**Table 2.6** (continued)

| S. No. | Maintenance | Frequency |
|---|---|---|
| 7. | Validate operation and calibration of temperature, pressure switches, and gauges | Once every five years |
| 8. | Check heater operation, tightness of terminal, linkages, screws, bolts | Once every five years |
| 9. | Latch, linkage, and operating mechanism adjustments | Once every five years |
| 10. | Overhaul breaker with new seals, contacts, and nozzles | Once every 10th to 15th year or 4000 to 10,000 operations |
| 11. | Overhaul disconnect, grounding, and breaking switches | Once every 15 years or 5000 to 10,000 operations |

**Table 2.7** International practices, vacuum CB inspections, diagnostics, and maintenance

| S. No. | Inspections | Frequency |
|---|---|---|
| 1. | General visual inspection (looking for obvious defects) | Once per month |
| 2. | Visual check for leaks/contamination/moisture/loose connections | Once every six months to annually |
| 3. | Visual check of terminations | Once every six months |
| 4. | Functional test | Annually |
| 5. | Record meter readings, check temperature, pressure switch operation, and calibrate gauges | Annually |
| 6. | Check foundations, grounds, paint | Annually |
| S. No. | Diagnostics | Frequency |
| 1. | Measurement of contact resistance | Annually to once every three years |
| 2. | Measurement of switching time and simultaneity of the phases. Lubricate moving parts at this time | Annually to once every three years |
| 3. | Measure spring charging time for spring-operated breakers | Once every three years |
| 4. | Partial discharge measurements | Once every six years |
| 5. | Vacuum tests in order to determine dielectric properties | Once every six years |
| 6. | Power factor, AC high potential | Once every five years |
| S. No. | Maintenance | Frequency |
| 1. | Lube, clean, adjust, align control mechanisms | Annually |

**Table 2.8** International practices, MV transformer inspections, diagnostics, and maintenance

| S. No. | Inspections | Frequency |
|---|---|---|
| 1. | General visual inspections (looking for obvious defects) | Monthly |
| 2. | Contamination on bushings/fan (grates) | Monthly |
| 3. | Corrosion fans (ONAF)/secondary equipment cabinet/ transformer | Monthly |
| 4. | Oil leaks | Monthly |
| 5. | Condition of silica gel desiccant | Monthly |
| 6. | Check for loose connections | Monthly |
| 7. | Visual inspection of HV and LV cable end boxes for oil leaks | Monthly |
| 8. | Monitor transformer temperatures (top oil, thermal image), load, and ambient temperature, and check for unusual coherence between these factors | Continuously (online) |
| 9. | Infrared inspection | Annually |
| S. No. | Diagnostics | Frequency |
| 1. | Dissolved gas-in-oil analysis (DGA) | Annually to once every three years |
| 2. | Oil quality tests. Includes:<br>a. Breakdown voltage<br>b. Neutralization value<br>c. Dielectric dissipation factor<br>d. Interfacial tension | Once every four to six years |
| 3. | Insulation resistance test | Once every eight years |
| 4. | Insulation power factor test | Once every eight years |
| 5. | Winding excitation test | Once every eight years |
| 6. | Bushing power factor test | Once every eight years |
| 7. | Turns ratio test | Once every eight years |
| S. No. | Maintenance | Frequency |
| 1. | Dust cleaning | Annually |
| 2. | Functional tests of all protection and instrumentation, for example:<br>a. Bucholz relay<br>b. Thermal image (indication winding temp)<br>c. Level indicators<br>d. Indicators (needle) and contacts (needle activates contact reaching certain threshold)<br>e. Fans (functional and timing)<br>f. Pressure relay | Annually |

(continued)

**Table 2.8** (continued)

| S. No. | Inspections | Frequency |
|---|---|---|
| 3. | Off load test—complete range tap changer in order to clean the contacts<br>a. In the case of an existing transformer: only OLTC (on load tap changer), do not operate DETC (de-energized tap changer) due to possibility of failure (cause: carbon on contacts)<br>b. In the case of a new transformer: both OLTC and DETC, but it should be done every year | At least during scheduled shutdowns |
| 4. | Inspect the diverter switch | Every 40,000 operations (typically). Number of operations depends on LTC manufacturer |

## 2.11 Maintenance and Inspection Roadmap Recommendations

Maintenance is the largest internally controllable cost. At the same time, maintenance is a key process in effective asset management [16]. The utility is in the process of improving its maintenance program and is investigating options to establish an effective maintenance program in a time- desired frame. A maintenance optimization roadmap is recommended based on traditional reliability-centered maintenance with elements of modern asset management (AM) and performance-based maintenance (PBM). Traditional RCM is focused on balancing the cost of maintenance and PBM reinforces the need for establishing processes for effective performance management through a "living" maintenance optimization program. This is a data-driven process and a suggested roadmap addresses the need for data management and support through a computerized maintenance management system.

The maintenance optimization roadmap would suggest number of individual steps in establishing effective maintenance consistent with modern asset and support reliability management as part of overall company performance management. The roadmap is organized in sections addressing the need for strategic decisions and suggesting project organization, planning, and prioritization. It includes pilot projects, technology transfer, and RCM studies. Once a maintenance plan is developed, the roadmap requires effective plan execution through proper work management and prioritization, resource optimization, and performance management. Steps to ensure effective data collection, management, and analysis are suggested together with installation of effective IT infrastructure.

The roadmap proposes maintenance optimization through the optimized use of internal resources in combination with external support to ensure cost-effective

**Table 2.9** International practices, station (power) transformer inspections, diagnostics, and maintenance

| S. No. | Inspections | Frequency |
|---|---|---|
| 1. | Overall visual inspections of connections, corrosion, oil levels, and so on | Monthly |
| 2. | Visual check of cable end boxes | Monthly |
| 3. | Visual check of breather; replacement if broken | Monthly |
| 4. | Visual check of silica gel desiccant; replacement if applicable | Monthly |
| 5. | Infrared inspections | Annually |
| 6. | Heat exchanger inspections | Annually |
| 7. | Conservator and bladder inspections | Annually |
| S. No. | Diagnostics | Frequency |
| 1. | Dissolved gas-in-oil analysis (DGA) | Annually to once every three years |
| 2. | Oil quality tests. Includes:<br>a. Breakdown voltage<br>b. Neutralization value<br>c. Dielectric dissipation factor<br>d. Interfacial tension | Once every four to six years |
| 3. | Insulation resistance test | Once every eight years |
| 4. | Insulation power factor test | Once every eight years |
| 5. | Winding excitation test | Once every eight years |
| 6. | Bushing power factor test | Once every eight years |
| 7. | Turns ratio test | Once every eight years |
| S. No. | Maintenance | Frequency |
| 1. | Dust cleaning | Annually |
| 2. | Functional tests of all protection and instrumentation, for example:<br>a. Bucholz relay<br>b. Thermal image (indication winding temp)<br>c. Level indicators<br>d. Indicators (needle) and contacts (needle activates contact reaching certain threshold)<br>e. Fans (functional and timing)<br>f. Pressure relay | Annually |
| 3. | Off load test—complete range tap changer in order to clean the contacts<br>a. In the case of an existing transformer: only OLTC (on load tap changer), do not operate DETC (de-energized tap changer) due to possibility of failure (cause: carbon on contacts)<br>b. In the case of a new transformer: both OLTC and DETC, but it should be done every year | During scheduled shutdowns, but at least |
| 4. | Inspect the diverter switch | Every 40,000 operations (typically). Number of operations depends on LTC manufacturer |

**Table 2.10** International practices, distribution transformer inspections, diagnostics, and maintenance

| S. No. | Inspections | Frequency |
|---|---|---|
| 1. | Overall visual inspections of connections, corrosion, oil levels, and so on | Annually |
| 2. | Inspection for physical damage | Annually |
| 3. | Visual check cable end boxes | Annually |
| 4. | Visual check of breather; replacement if broken | Annually |
| 5. | Visual check silica gel, replacement if applicable | Annually |
| S. No. | Diagnostics | Frequency |
| 1. | Oil sampling = Dissolved Gas Analysis (DGA) | Once every two years. The DGA should be performed only on the critical transformers (important customers) or in the case of an indication of a possible fault |
| 2. | Partial discharges, when discharges are detected oil samples (DGA) should be taken in order to determine the composition of key cases and localize the cause of the problem. In the case of abnormal situations the oil should be monitored on a regular basis in order to analyze a possible trend and determine further actions. In the case of a short-circuit current the transformer should also be measured to cheek for abnormalities. Thermographic measurement should be undertaken | Once every six years |
| 3. | Dust cleaning | Annually |

**Table 2.11** International practices, instrument transformer inspections, diagnostics, and maintenance

| S. No. | Inspections | Frequency |
|---|---|---|
| 1. | Visual inspection | Annually |
| 2. | Infrared scan | Annually |
| S. No. | Diagnostics | Frequency |
| 1. | Power factor test | Once every five years |
| 2. | Burden measurements | Once even, five years |

implementation and technology transfer with high quality of deliverables. The proposed implementation of computerized maintenance management and upgrade of IT infrastructure is the highest "hard" cost of the project. For the reason of cost and ROI management and because the same infrastructure should be used for companywide asset management and overall performance (including reliability) management the roadmap suggests a two-step, companywide approach in investing

**Table 2.12** International practices, capacitor bank inspections, diagnostics, and maintenance

| S. No. | Inspections | Frequency |
|---|---|---|
| 1. | Visual inspection | Monthly |
| 2. | Detailed visual inspection. Check for oil leaks, bulging of the capacitors, signs of overheating, arcing point, and terminations | Annually |
| 3. | Measure temperature (summertime, representative temperatures) in order to verify whether the temperature is exceeding the limit of the rated temperature class. If the temperature class is exceeded, possibilities for (natural) cooling should be examined | Annually |
| S. No. | Diagnostics | Frequency |
| 1. | Capacitance measurement of complete bank | Annually |
| 2. | Upon deviation, measure value of each unit | Annually |
| 3. | At deviation from manufacturer tolerances, replacement of individual unit(s) | Annually |
| 4. | Partial discharge measurements | Annually |
| 5. | Thermographic measurements | Annually |

**Table 2.13** International practices, MV switchgear and control panel inspections, diagnostics, and maintenance

| S. No. | Inspections | Frequency |
|---|---|---|
| 1. | General visual inspection; in the case of digital relay check El n0 and ET i z 0 if applicable (indication of status of current and voltage transformer and wiring) | Monthly |
| 2. | Electromechanical and static relays: functional check and primary/secondary injection; replacement if applicable | Annually |
| 3. | Check parameter adjustment (mainly caused by capacitors) | Annually |
| S. No. | Maintenance | Frequency |
| 1. | Digital relays, due to self-check ability of relay: functional check and primary/secondary injection; replacement if applicable | Once every three years |

in infrastructure upgrade with the first step being detailed specification and requirements development taking into account companywide asset and performance management needs.

**Maintenance and Inspection Roadmap Conclusions**

A utility has a very good opportunity to achieve desired reliability transformations. This goal is achievable through a coordinated strategy at the corporate level and continued enthusiastic support from top management. Maintenance improvement, as discussed in the maintenance optimization roadmap, is an initiative that needs companywide coordination. Investment in a maintenance optimization project is an investment in the company's overall performance improvement, including reliability.

**Table 2.14** International practices, DC system: Batteries and battery charger inspections, diagnostics, and maintenance

| *Battery* | | |
|---|---|---|
| S. No. | Inspections | Frequency |
| 1 | General visual inspection | Monthly |
| 2 | Check liquid level and refill if applicable | Monthly |
| S. No. | Diagnostics | Frequency |
| 1 | In the case of open batteries check PH value | Annually |
| 2 | Voltage measurement | Annually |
| 3 | No break test, cell replacement if applicable | Once every two years |
| S. No. | Maintenance | Frequency |
| 1 | Check tightness and oxidation of connections and maintain if applicable | Once every two years |
| *Rectifier* | | |
| S. No. | Inspections | Frequency |
| 1 | Visual inspection | Once every two years |
| 2 | Check functionality of the current rectifier, both float and boost positions, if applicable | Once every two years |
| 3 | Check indication lights | Once every two years |
| 4 | Check for contamination, cracks, and arcing points | Once every two years |
| 5 | Check indicators (voltage, current) | Once every two years |

**Table 2.15** International practices, feeder pillar inspections, diagnostics, and maintenance

| S. No. | Inspections | | Frequency |
|---|---|---|---|
| 1. | Inspection for physical damage; repair or painting if applicable | | Annually |
| S. No. | Diagnostics | | Frequency |
| 1. | Check maximum current indicator and compare it with ratings | | Annually |
| 2. | Infrared measurement | | Once every three years |
| S. No. | Maintenance | Frequency | |
| | | Normal Environmental Conditions | High Contamination Areas |
| 1. | Cleaning | Once every three years | Once every three years |
| 2. | Tightening of loose connections | Once every three years | Once every three years |

**Table 2.16** International practices, ring main unit inspections, diagnostics, and maintenance

| S. No. | Inspections | Frequency |
|---|---|---|
| 1. | Visual check of cable terminations | Every six months |
| 2. | Check operating mechanism for corrosion and/or contamination. Clean and lubricate if applicable | Annually |
| 3. | Inspect for physical damage/corrosion | Annually |
| S. No. | Diagnostics | Frequency |
| 4. | Infrared measurement | Every three years |
| S. No. | Maintenance | Frequency |
| 5. | Cleaning | Every three months; only for aggressive environmental conditions (high contamination areas) |

Maintenance is the largest internally controllable cost and is a key process in effective asset management. It is recommended to have a maintenance optimization roadmap based on traditional reliability-centered maintenance in combination with elements of modern asset management and performance-based maintenance. Traditional RCM is focused on balancing the cost of maintenance and reliability. The PBM reinforces the need for establishing processes for effective performance management through a "living" maintenance optimization program. As stated above, this is a data-driven process and the suggested roadmap addresses the need for data management and supporting IT infrastructure through a computerized maintenance management system.

The important element in considering cost and benefit of the project is to realize that most of the work could be done using internal resources. This is important not just from the cost-saving perspective, but is also a good and effective way of technology transfer. This approach helps to mobilize internal resources from the beginning of the project, and is an effective way to minimize implementation time and ensure internal ownership of the project.

Cost of the investment in performance management infrastructure, data standardization, and data collection should be considered a companywide cost and a project of long-term significance with wide-ranging benefits.

If managed properly, investment in the maintenance optimization project, processes, and infrastructure suggested in the roadmap will take full advantage of industry best practices and "lessons learned" and turn the existing utility's position of "being late" into the advantage of "being ahead" through effective implementation of the latest concepts and tools without being held down by old historical processes and infrastructure [17].

The recommendations related to the maintenance optimization roadmap provide a scenario of full implementation as the desired goal for transformation from time-based to maintenance to RCM. The goal of a predefined period in essence is achieved in the proposed roadmap through the implementation of the most critical RCM projects within the desired timeframe and continuous RCM implementation across less critical components. If required, a shorter implementation schedule is possible by assigning high-priority status to the project, shortening the implementation cycle through accelerated installation of required infrastructure, and allocation of a higher percentage of available internal labor resources to the project technology development and implementation.

## 2.12 Maintenance Management

A proper maintenance program is formulated for all distribution systems because they are distinct in their configuration, load, and types of installed equipment. The present trend is to move ahead from traditional preventive maintenance methods to PDM using CBM (condition-based monitoring tools) and finally targeting RCM.

**Need for Maintenance Program**

Electrical equipment deterioration begins when new equipment is installed in a system; it occurs at a standard rate, however, equipment failure is not predictable. When the equipment is not properly maintained or unchecked, equipment performance decay may occur, leading to fault or failure of the equipment, particularly a performance decay rate increase in a hostile environment condition, overload, or severe duty cycle. Apart from the ageing factor, other reasons for failure of equipment may be monitored and corrected through such a program. At the time of scheduling the program for maintenance, the following issues should be taken into consideration.

- Worker safety
- Equipment loss
- Production economics

The vital elements of a thriving maintenance program are:

- Deploying qualified personnel and responsibilities
- Detailed electrical equipment survey and analysis for maintenance
- Necessities and importance
- Preplanned repetitive inspections and tests
- Exact investigation of inspection and test reports can lead to proper corrective action
- Complete and concise records

**Emergency Procedure**

Step-by-step procedures should be adopted in the event of an emergency, including secure stoppage or startup of equipment and systems. Well-maintained safety equipment is critical and necessary when working on or near live electrical equipment.

**Identification of Critical Equipment**

After the fault, failure equipment is considered critical, even when the equipment under complete control during normal operation may be a serious risk to maintenance personnel, property, or the product. The criticality of a machine can be determined based on the previous experience and knowledge of maintenance personnel. An intact system may be critical by its very nature. Examples of critical systems are:

(a) Emergency lighting and power
(b) Fire alarm and pumps
(c) Certain communication systems

Several parts of a system may also be critical because of the role of the utilization equipment and the related hardware.

**Scheduling**

The scheduled checkups and tests are the essential part of an effective maintenance program. Maintenance reveals the condition of equipment, and tells us what kinds of maintenance are required and whether the equipment will continue to operate properly until the next schedule. Factors to be taken care of are:

- Surroundings where the electrical equipment is located
- Condition of load
- Equipment history
- Frequency of inspection

**Effective Maintenance Program Methods and Procedures**

These must include the connections between individual components. Neglecting the system's interconnections and operation together may result in unexpected problems. The system procedures should contain:

(a) An assortment of forms for use in the plant and the field.
(b) Each piece of equipment should have a set of procedures detailing the special tools, materials, and necessary equipment as well as an estimation of the time to perform the work, references to suitable technical manuals, documentation of preceding work performed, items of particular consideration, precautions, uncommon events, and so on.
(c) Safety procedures to be inculcated.
(d) The operating personnel should follow the schedule for plant inspection as per frequencies of schedule, accessibility, readiness of auxiliary equipment, shutdowns as per schedule, projected inspections and tests, readiness of standby parts, and special test equipment.
(e) Record maintenance to assess results.
(f) Emergency procedures that include training in emergency situations.

## 2.13 New Technologies in Power System Control Operation

**Supervisory Control and Data Acquisition (SCADA)**

Since the inception of digital computer use for applications apart from data processing in offices SCADA systems have been used in various industries for many years. The introduction of microprocessor technology has extended the relevance of SCADA technology through the more sophisticated concept of the distributed control system (DCS).

**SCADA Systems History**: The SCADA system is mostly used to control and monitor the generating station output and the HV transmission systems. In recent years, the SCADA system is more cost effective to monitor and control electrical distribution systems particularly when the system voltage is less than 15 kV.

**SCADA System for Electrical Distribution**: The financial validation for a devoted SCADA system is frequently tricky. The attention of datapoints is typically relatively low. The configuration at any one site is unique. The SCADA system is mainly used to monitor voltage, current, power flow of major circuits, and the monitoring and control of circuit breakers and generation units. The configuration of an electrical system can be altered remotely by the closing or opening of CB and switches. Faults on the system are detected right away, and alarm conditions and operation of protective devices are reported to a central location. The SCADA system stores and accumulates the necessary information for load shedding and energy conservation, and is readily obtainable for the management of engineering personnel. Accumulation of data can ease maintenance, troubleshooting, and support emergency planning.

**Typical SCADA System Configuration**: Modern SCADA systems consist of a minimum of three or over a hundred central stations; each central station normally has one or more computer consoles and two or more central computers are in surplus configuration. Each central station contains the computer, communications equipment, and power supplies; integral or separate data and alarm printers; and configuration-dependent terminal strips (IED-intelligent electronic devices). These are essential for communicating with remote equipment. The IEDs used in the digitalization of distribution substations include security, automation, control, and communications. They gather and generate significant analytical data by monitoring the devices and surroundings that the IEDs protect. Each main device of a substation has at least one associated IED receiving voltage, current, status, and other signals that are put to use to make precise diagnoses in real-time.

The IEC 61850 standard establishes the functionality of the vertical and horizontal communication protocols, enabling interoperability between the systems and quick exchange of multiple types of messages among the protection, control, supervision, and measurement system equipment of the substations (the IEDs). Power distribution substations designed to use IEC 61850 can be optimized by using messages designed for SCADA, realtime data exchange, and collection of equipment monitoring information, significantly reducing equipment used for protection, control, measurement, and automation. The computer consoles have one or more cathode ray tube (CRT) display devices, with the facility of operator communication: a touchtype keyboard, a touchscreen, or a light wand. The actual CRT displays are generated by specially written computer software consisting of schematic multicolor representations of a variety of operating systems.

Electrical power systems are frequently represented in a one-line diagram with some geographical or physical orientation of assorted equipment. The system operating buses are displayed in diverse colors to reflect the dissimilar operating voltages. CB status is indicated by red or green lights according to their closed or open status, respectively. Bus power flows, voltages, and currents are displayed next to the bus identification information. Transformers, circuit breakers, disconnect switches, major circuits, and major utilization equipment are usually illustrated on the display. The display may be broken down into supplementary detailed screens to exemplify the particulars of a complex system. The control conditions for generating units if any should be displayed. A suitable communication network is placed between the central control room and remote stations (PLCC, optical fiber cable, etc.).

**Control Circuits and Devices**

Microprocessor controls are one of the latest technologies to be applied to the electrical distribution system. Microprocessors are being put to use for applications as follows.

(a) Protective relaying and tripping functions in circuit breakers and fuse like switching devices
(b) Electronic meters that provide all of the voltage: current, power, energy consumption, demand, power factor, frequency

Microprocessor-based protection modules are being installed in molded case circuit breakers and low voltage power circuit breakers. This aids in controlling the operation of the direct acting trip units. These trip units are used to provide long time, short time, instantaneous, and ground fault over current protection. They also take care of under-voltage protection. Furthermore, microprocessor-based protective relays are also put to use for replacing traditional electromechanical protective relays generally used for low and medium voltage switchgear installations. The novel devices proffer better protection of equipment by allowing more precise protection settings paralleling equipment needs. The latest devices also present improved troubleshooting diagnostics, online test features, and communication capabilities. This leads to remote trouble reporting. These protective devices are being outfitted with metering capabilities that may allow removal of separate voltmeters, ammeters, and wattmeters, which are frequently used on feeder circuits and utilization circuits for providing operating load information. Electronic metering devices are now obtainable in one package to substitute all independent meters.

## 2.14 New Systems and IT Interventions

With the introduction of IT-enabled systems in distribution business operations, many new types of software have been introduced. To name a few of them:

- GIS mapping and engineering
- Metering, billing, and collection
- Trouble call management (call center)
- Project management
- ERP

Today these may be working in silos. A need has therefore arisen to have EAI (enterprise application integration) in place so all activities of the distribution utility are carried out in a harmonious way.

**Operating Responsibilities and Organizational Relationships**
A detailed work breakdown structure is required to be drawn to divide the responsibilities across various divisional units and personnel working there, such as

- Planning and project
- Engineering and procurement
- Operation and maintenance
- Protection, metering, and testing.

## References

1. Belighter Charles S et al (1982) Foundation of optimization, Prentice-Hall of India, New Delhi
2. Masu dEnver (1974) An interactive procedure for sizing and timing of distribution substation using optimization techniques. IEEE Trans 93:1281–1286, Sept–Oct
3. Renationalization of Primary Distribution Network, Technical Report No.137, CPRI Bangalore, May 1996
4. Electric Power International (1981) ERA technology, London, pp. A-19–A-21
5. Pabla AS (1984) Average power distribution system, Proceeding of Power India Tech. in Transition Conference, NPC, New Delhi, March 7–8
6. Standard Handbook on Electrical Engineering, 12th Edition (1987) pp. 17/3–7
7. Petterson L (1990) Estimation of the remaining service life of power transformer and their insulation. Electrica, CIGRA, 133:65–71, Dec 1990
8. Kothari DP, Dhillon JS (2004) Power system optimization. Prentice-Hall of India, New delhi
9. Jennings Bench B, Richard D, Miller & James E, Wheeler (1982) Distribution system integrated voltage & reactive power control. IEEE Trans PAS-101, Feb, pp 284–286
10. Lynn P (2004) Power system load flow analysis. TATA McGraw Hill, New York
11. Wallach Y (1986) Calculations and programs for power system networks. Prentice-Hall, New Jersey
12. Central (2007) Electricity Authority (Grid Connectivity Regulations), New Delhi

13. Lee WH (1997) Power distribution planning. Reference Book Marcell Dekker Inc, New York, p 209
14. Stall HG (1989) Least-cost electric utility. Willey, New York, p 74
15. Geographic Information System Planning for power system planning, United Nation Publication, June 1997
16. Grisby LL (1998) The electric power engineering, Handbook, pp 6–60, 13-4
17. National Electricity Plan 2004, CEA, New Delhi (Chap. 7)

# Chapter 3
# Best Practices in Operation and Maintenance of Subtransmission and Distribution Lines

## 3.1 Description of Distribution Network

In a power system the distribution network is the final stage, which includes carrying electricity from the transmission system and also delivering the electricity to the end users (Fig. 3.1). The components of this network are power lines (medium voltage below 66 kV), substations, distribution transformers, meters, and low-voltage distribution wings (below 1 kV). It begins with the primary circuit leaving the substation and ends as the secondary service enters the customer's meter socket [1]. The voltage varies from 33 kV to 230 V depending on the utility standards, distance, and load to be served.

Distribution system conductors are carried on overhead poles; in high-density areas underground cables are used. Three-phase four-wire systems are used to serve electricity for urban and suburban area load customers, whereas for rural areas single-phase systems are used. Generally, a distribution transformer is used to feed domestic customers. The distribution transformer reduces the voltage to the relatively low level used by lightning and the interior wiring system. Large commercial customers are fed directly from distribution voltages.

The transformer can be set up in two ways: either pole-mounted or also mounted on the ground in a protective covering. A pole-mounted transformer can serve only one transformer in the case of rural areas, and in builtup areas, more than one customer may also be connected. A secondary network, in which many transformers feed into a common bus at utilization voltage, may be formed in very dense city areas. Every customer has electrical service or a service drop connection and a meter for billing purposes. (A few small loads such as yard lights are too small for a meter and are billed only at a monthly rate.)

For the customer's system and equipment owned by the utility, a ground connection to local earth is normally provided. The customer's system is connected to the ground in order to limit the voltage that can increase when high-voltage conductors fall on lower-voltage conductors, or even in case of failure within the



J. Raja et al., *Practices in Power System Management in India*,
Power Systems, https://doi.org/10.1007/978-981-10-2972-1_3

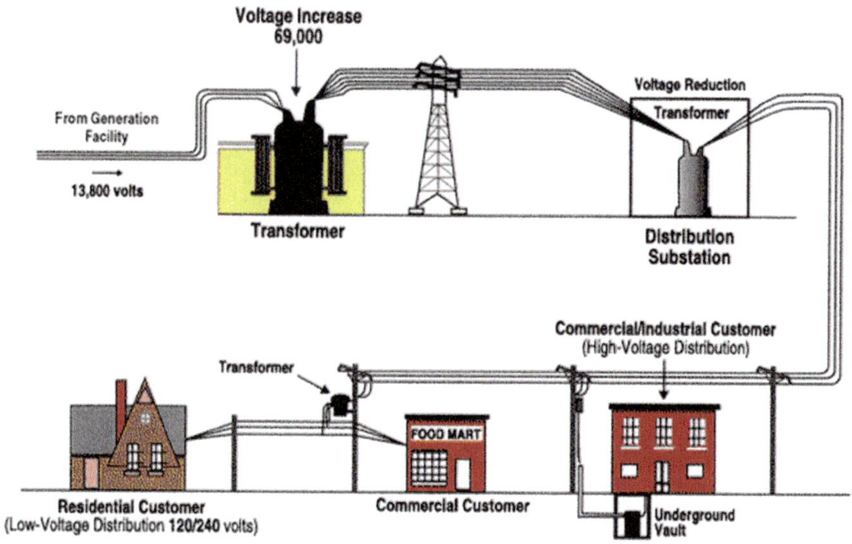


**Fig. 3.1** Structure of distribution system

distribution transformer. The risk of electric shock can be minimized when all conductive objects are connected together to a common earth grounding system. But the presence of multiple connections between utility ground and customer ground may lead to stray voltage problems such as customer piping, swimming pools, or any other equipment that may develop objectionable voltages [2, 3]. Because problems often develop from places separated from the customer's premises, these are difficult to solve.

Generally, a distribution network contains the following tools.

## 3.2 Poles and Tower Structures

The tower can be supported by narrow-based lattice towers with a fully galvanized structure according to the requirements. Poles are used for 33, 11 kV, and low transmission lines (below 500 V). There are many kinds of poles such as precast concrete (PCC), prestressed cement concrete (PSCC), rolled steel joist, rail, or tubular steel (Table 3.1). The standards confirmed by the poles should be IS 785, IS 1322, IS 1678, IS 2713. Cement poles are favored in plain areas. In designing poles and towers in hilly areas, appropriate snow or ice loading should be taken into consideration.

Specially designed poles/lattice towers are used in locations involving long spans or higher clearances owing to the crossing of power or communication lines or railway line crossings. If the angle of deviation is more than 10°, double-pole structures are used. The factors that should be considered for determining the height

**Table 3.1** Types of structure by function

| | |
|---|---|
| **Wood pylons**: A straight trunk impregnated with tar is usually used with conductor cables on the top and carries one or more crossbeams for support pylons. Anchor pylons require higher force thus constructions looking like V or A are used. Wood pylons are used only for LV lines generally and height of pylons is also limited because of restricted height of trees available |  |
| **Concrete poles**: An electricity pole made from reinforced concrete is manufactured at the factory and put up at the powerline's right of way. These are prefabricated and used up to 60 m in height |  |
| **Steel rail**: These pole structures are high-tension carrying structures and are normally used at corners or turning points. Generally these rail prefabricated structures are installed with concreting at ground |  |
| **Steel tube**: This pole is normally used for street light connections and is manufactured from steel tubes. They are normally assembled at the factory and set up on the powerline's right of way with a crane |  |
| **Lattice steel pole**: This electricity structure consists of a steel framework construction and is used for powerlines of all voltages. They are normally assembled from individual parts at the place where it is to be erected |  |

of the pole above ground level, length of the pole below ground level, and load are wind zone, terrain, topography, and statutory clearances required to be maintained. All these factors are required to be confirmed by Indian standards (IS) [4].

Depending on the type of line there can be a variety of shapes for the structure of overhead lines. A simple structure is made up of wood poles straight in the earth that carry one or more cross-arm beams in order to carry the conductors. It can also be constructed as an "armless" design with conductors reinforced on insulators on the sides of the poles. In urban areas, tubular steel poles are generally used. Lattice-type steel towers or pylons are used to carry high-voltage lines. In remote areas, aluminum towers are used and sometimes concrete poles may be used. Due to higher costs, poles made of reinforced plastics are restricted in use.

Each structure must be designed for the loads imposed on it such as conductors, wind, ice, and so on; however, this is a well-known design that answers to specific local and national regulation. A large transmission line project can have various kinds of towers such as one with "tangent" (suspension or line) towers meant for various locations and more heavily built towers used for turning the line on an angle or dead-ending (terminating) which can either be a line or important river/road intersection (Table 3.2). There are semiflexible types of structures that can rely on the weight of conductors to be balanced on either side of each tower and these depend on the design criteria for the particular line. There are more rigid structures that are made so they remain standing even when conductors are broken. There may be cascading tower failures and their scale can be limited by installing such structures at gaps in power lines.

**Table 3.2** Type of PCC poles for angle and dead-end locations

| S. No. | Location and type of support | Suitability of different PCC pole designs in Western part (WP) zones | | | |
|---|---|---|---|---|---|
| | | 50 kg/m$^2$ | 75 kg/m$^2$ (using 140 kg WL poles in normal location) | 75 kg/m$^2$ (using 200 kg WL poles in normal location) | 100 kg/m$^2$ |
| 1 | Straight runs (tangent locations): single-pole arrangement | A | A | B | B |
| 2 | 0°–10° angle location: single-pole arrangement | A | A | B | B |
| 3 | 10°–30° angle location: single-pole arrangement | A | A | B | B |
| 4 | 10°–30° angle location: double-pole arrangement | A | A | B | B |
| 5 | 30°–60° angle location: single-pole arrangement | A–A | A–A | B–B | B–B |
| 6 | 3°–60° angle location: four-pole arrangement (H-type) | A | A | B | B |
| 7 | 60°–90° angle location: double-pole arrangement | A | A | B | B |
| 8 | Dead-end location: double- pole arrangement (H-type) | A | A | B | B |
| 9 | Distribution transformer location: double-pole arrangement | C | C | D | D |

### 3.2.1 Selection Criteria of PCC Poles per Requirement of Electric Connection (REC)

See (Table 3.2).

## 3.3 Line Span

In overhead lines the conductors are supported by towers/poles (Tables 3.3 and 3.4). The conductors are pulled and stringing effected [5]. When supported this way a conductor will dip under its own weight called "sag." The distance between adjunct supporting structures is called the "span." The conductor's mechanical loading is due to its own weight, weight of ice, wind load, and so on. Under varying weather conditions of ambient temperature, the conductor's tension should not

**Table 3.3** Permissible spans for 11 kV lines

| Conductor | Working load of supports (Kg) | Max. permissible span on a wind pressure zone | | |
|---|---|---|---|---|
| | | 50 kg/m$^2$ | 75 kg/m$^2$ | 100 kg/m$^2$ |
| 1 | 2 | 3 | 4 | 5 |
| 11 kV line, cross country, on 7.5 m supports—conductor formation and clearance, per REC construction start A-4 | | | | |
| ACSR 7/2.11 mm | 140 | 107 | 107 | NR |
| ACSR 7/2.11 mm | 200 | NR | 107 | 107 |
| ACSR 7/2.59 mm | 140 | 107 | 90 | NR |
| ACSR 7/2.59 mm | 200 | NR | 107 | 97.5 |
| ACSR 7/3.35 mm | 140 | 107 | 69 | NR |
| ACSR 7/3.35 mm | 200 | NR | 106 | 75 |
| 11 kV line, cross country, on 8.0 m supports—conductor formation and clearance, per REC construction start A-5<br>Earth-wire 4 mm (550–900 MPa quality) | | | | |
| ACSR 7/2.11 mm | 140 | 107 | 91 | NR |
| ACSR 7/2.11 mm | 200 | NR | 107 | 99.5 |
| ACSR 7/2.59 mm | 140 | 107 | 76.5 | NR |
| ACSR 7/2.59 mm | 200 | NR | 107 | 83 |
| ACSR 7/3.35 mm | 140 | 99 | 60.5 | NR |
| ACSR 7/3.35 mm | 200 | NR | 93.5 | 66 |
| 11 kV line, cross country, on 8.0 m supports—without earth-wire, conductor formation, and clearance per REC construction standard A-5 (excluding earth-wire) | | | | |
| ACSR 7/2.11 mm | 140 | 107 | 107 | NR |
| ACSR 7/2.11 mm | 200 | NR | 107 | 107 |
| ACSR 7/2.59 mm | 140 | 107 | 87.5 | NR |
| ACSR 7/2.59 mm | 200 | NR | 107 | 95 |
| ACSR 7/3.35 mm | 140 | 107 | 67.5 | NR |
| ACSR 7/3.35 mm | 200 | NR | 104 | 73.5 |

**Table 3.4** Maximum permissible spans for three-phase LT lines

| Conductor | Working load of supports (Kg) | Max. permissible span on a wind pressure zone | | |
|---|---|---|---|---|
| | | 50 kg/m$^2$ | 75 kg/m$^2$ | 100 kg/m$^2$ |
| 1 | 2 | 3 | 4 | 5 |
| 415/240 V lines, cross country, on 7.5 M support—conductor formation and clearance per REC construction standard B-12 (s-phase, four-wire horizontal) | | | | |
| AAC 7/2.11 mm | 140 | 81 | 73 | NR |
| AAC7/2.11 mm | 200 | NR | 73 | 66 |
| ACSR 7/3.10 mm | 140 | 80 | 61 | NR |
| ACSR 7/3.10 mm | 200 | NR | 74.5 | 66 |
| ACSR 7/2.11 mm | 140 | 107 | 83 | Nr |
| ACSR 7/2.11 mm | 200 | NR | 107 | 90.5 |
| ACSR 7/2.59 mm | 140 | 107 | 71 | NR |
| ACSR 7/2.59 mm | 200 | NR | 107 | 77 |
| ACSR 7/3.35 mm | 140 | 93 | 57.5 | NR |
| ACSR 7/3.35 mm | 200 | NR | 88 | 62.5 |
| 415/240 V lines, cross country, on 8 M support—conductor formation and clearance per REC construction standard B-13 (three-phase, four-wire horizontal) | | | | |
| AAC 7/2.21 mm | 140 | 81 | 73 | NR |
| AAC 7/2.21 mm | 200 | NR | 73 | 66 |
| AAC 7/3.10 mm | 140 | 80 | 61 | NR |
| AAC 7/3.10 mm | 200 | NR | 74.5 | 66 |
| ACSR 7/2.11 mm | 140 | 107 | 83 | NR |
| ACSR 7/2.11 mm | 200 | NR | 107 | 90.5 |
| ACSR 7/2.59 mm | 140 | 107 | 71 | NR |
| ACSR 7/2.59 mm | 200 | NR | 107 | 77 |
| ACSR 7/3.35 mm | 140 | 93 | 57.5 | NR |
| ACSR 7/3.35 mm | 200 | NR | 88 | 62.5 |

exceed the permissible limit, that is, breaking strength of the conductor/safety factor (2.0–2.5). Knowledge of the maximum sag/line span calculation is essential for designing lines with adequate ground clearance. The line span is decided by taking into account factors such as topography, wind pressure, kind of support, conductor configuration, and ultimate tensile strength.

The span range is designated by IS-5613.

There should be a uniform span that is to be maintained between two successive pole structures as soon as possible. A pole should be placed on the roadside if a road is intersected at midspan when the line is built. When another power line is crossed, the lower voltage line should be below specified. The lower line should be crossed in the midspan of the upper line [6, 7]. In order to maintain the proper ground clearance at the middle of the span, shorter poles can be made when poles are placed at high places. Normally, placement of poles should be avoided along the edges, cuts, or embankments of creeks and streams.

### 3.3.1 *Permissible Line Span per REC Standards*

See (Table 3.4).

## 3.4 Overhead Distribution Lines

In terms of functions, transmission lines and overhead distribution lines may be similar but these two may differ in terms of their construction. In terms of voltage and distance, transmission lines are made to carry high voltages traveling longer distances, but overhead distribution lines are constructed to carry medium voltages traveling distances that depend on the size of its extent or on the utility's discretion [8].

Electrical power starts at the power distribution substation and goes to different areas with the help of distribution lines in the system. Distribution lines connected to the high-voltage side of the distribution transformer are referred to as primary distribution lines or primaries. The secondaries, connected on the low-voltage side of the distribution transformer, are known as secondary distribution lines.

### 3.4.1 *Preventive Maintenance of Overhead Lines*

See (Table 3.5).

## 3.5 Cables

Cables are nothing but conductors covered with a layer of insulation (Fig. 3.2). Cables can have either one conductor or multiple conductors and may be of individually insulated type. There may also be multiple conductor cables with an external insulating outer covering or sheath. The materials used for making the conductor are electrolytic-grade high-conductivity annealed copper or aluminum. All cables generally have aluminum/copper as conductor materials and control cables have copper.

**Conductor Shield**

The conductor has a nonmetallic semiconducting shield that ensures a perfectly smooth profile and avoids stress concentration.

**Insulation** There is generally a specified system voltage for which cable insulation should be designed. The purpose of the manufacturing process is to ensure that insulation is free from voids. The insulation should also hold out against mechanical and thermal stress under steady- state as well as transient operating conditions.

**Table 3.5** Overhead line maintenance

| S. No. | Type of Inspection | Particulars of inspection | Maintenance |
|---|---|---|---|
| 1 | Line patrolling | 1. RC poles are broken | Replace the broken poles as a priority |
| | | 2. RS joist/rail poles, cross- arms are rusted | Clean the rusted poles or cross-arms by wire brush and paint with two coats of rustproof red oxide and two coats of synthetic enamel paint |
| | | 3. Stay set uprooted from ground/ stay wire broken | Excavate the pit, place the stay set properly and provide mass concreting up to ground level. Replace the broken stay wire with new one. Ensure the stay wire is in tension |
| | | 4. Coping of RS joints/rail poles near the ground is damaged | Provide proper cement coping with cement concrete |
| | | 5. Cattle rubbing on the pole at the ground | Provide barbed wire up to 1.5 m high on each pole to prevent cattle from rubbing against poles and also ensuring that unauthorized persons do not climb the live poles |
| | | 6. Kites and loose threads hanging on the overhead lines | Remove the kites, burn the threads, and clean the lines |
| 2 | Conditions of insulators | 1. Cobwebs, bird's nests, paper kites, and loose threads surrounding the insulators<br>2. Insulator found cracked/broken | Remove cobwebs, bird's nests, and threads; clean the insulators. Replace the damaged insulator with new insulator |
| 3 | Conditions of clamp/ fixtures | Clamps are loose. Bolts and nuts are missing. Fixtures are hanging loose | Replace the missing bolts and nuts. Tighten the clamps; firmly and permanently refix the fixtures by providing new clamps and bolts, nuts, and washers |
| 4 | Tree clearance | Tree branches are falling or spreading on the overhead lines, bushes and shrubs thickly surrounding the poles at ground level | Cut/remove such branches that are pressing the overhead lines. Remove the brushes and shrubs and keep the area near each pole clean |
| 5 | Inspection of earthing | Earth wires loosely hanging at the bottom of pole, earth- wire connection near the bottom of pole has come out. Earth lug found missing | Provide permanent connection. Replace missing or broken lugs and make the earth connections near the poles permanent by proper bolts, nuts, and washers |
| 6 | General testing | Megger used to test the overhead lines between phase-to-neutral, phase-to-earth and between neutral-to-earth | Check for any leakages and if found, patrol the line; check for faulty or broken insulators and replace them |
| 7 | | | |

(continued)

**Table 3.5** (continued)

| S. No. | Type of Inspection | Particulars of inspection | Maintenance |
|---|---|---|---|
| | Condition of ground wire | Guard wires loosely hanging on poles and guard cradles loosely hanging | String the wire and fix the cradles permanently |
| 8 | Logbook | | All above observations, rectifications, measurements, and so on to be recorded in logbook |

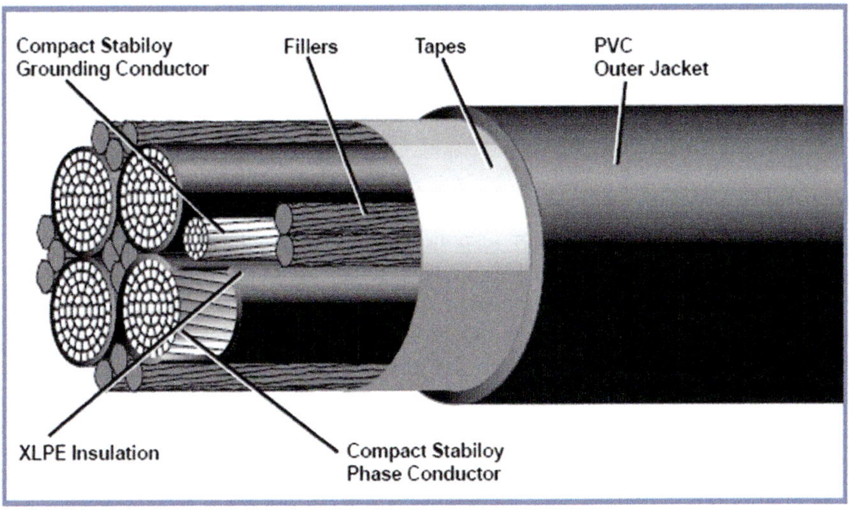


**Fig. 3.2** Cable internal composition

**Armoring** In single-core cables armoring is applied over the core insulation and in the case of multicore cable it is applied over the inner sheath. Depending upon the application, these cables can be armored or unarmored. The armor consists of galvanized round steel wire. In single-core cables to be used in an AC system, armoring with nonmagnetic material is desired [9]. For single- core cable to be used in a DC system, the cable should be armored with galvanized steel wire/strip.

**Sheath**

**Inner Sheath** In the case of cables consisting of two or more cores an inner sheath suitable to withstand the site conditions and the desired temperature is provided on the individual core and then surrounded by a common covering applied either by extrusion or wrapping of a filling material containing thermoplastic material. A PVC sheath should be extruded. The circularity of the cable should be maintained [10].

**Outer Sheath** This is extruded over the armor for armored cables and covers the inner sheath for unarmored cables. It is specially formulated heat-resistant black PVC extruded to form the outer sheath. The cable should also offer a specially formulated flame-retardant low smoke compound (FRLS) for the outer sheath. All must be given anti-termite treatment.

### *3.5.1 Preventive Maintenance of Underground Lines*

Table 3.6 shows underground line preventive maintenance.

## 3.6 Insulators

An insulator (Fig. 3.3) in the true sense is a material that completely resists the flow of electric charge and does not respond to the electric field. But, practically, a perfect insulator does not exist [11, 12]. Therefore dielectric materials having very high dielectric constants are regarded as insulators. In the atoms of insulating materials, electrons are strongly bonded. These are used as insulators or insulation in electrical equipment and perform the function of supporting or separating electrical conductors and do not allow current through them. The insulating supports used to attach electric power transmission wires to utility poles/pylons are also referred to by this term.

**Table 3.6** Preventive maintenance of underground lines

| Maintenance for underground cables | | | | |
|---|---|---|---|---|
| S. No. | Particulars | Type of installation | Inspection particulars | Inspection frequency |
| 1 | Cables of all grades: i.e., 1.1, 11, and 33 kV | Buried in trenches, ducts | Verification of conduction of cable | Once every half-year |
| 2 | Cables of all grades: i.e., 1.1, 11, and 33 kV | In soil | Whether there is water stagnation near the cable location | Rainy season |
| 3 | Cable termination for 1.1 kV grade cables | Open | Routine checks of condition | Once yearly |
| 4 | Cable terminations for 11 kV and above grade cables | Open | Routine checks of condition | Once yearly |
| 5 | Cable glands, lugs, and connections | Open | For loose joints, broken and missing glands, missing lugs, and loose connections. | Once every 3 months |

**Fig. 3.3** Different kinds of insulators

Glass, paper, or Teflon are some materials that are very good examples of electrical insulators because these materials may have lower bulk resistivity. However, a larger class of materials is "good enough" in insulating electrical wiring and cables. Some examples of such materials are rubberlike polymers and most kinds of plastics. For low to moderate voltages (hundreds, or even thousands, of volts), these materials are safe and practical insulators to use. Electric power transmission wires suspended in nature are generally bare and are insulated by only air. But when these wires enter a building, they aren't bare. Insulators are required at the points where they are supported by utility poles or pylons and where the wire enters buildings or electrical devices such as a transformer, circuit breaker, and the like. The hollow conductors that hold conductors in them are known as bushings [13]. It is necessary to use an insulating link by cranes working near such wires because of dangers produced by cranes that touch bare electric power transmission wires.

Excessive voltage can cause electrical breakdown of an insulator. This can take place in two ways:

- **Puncture voltage**: When the insulator is installed in its normal manner there is a voltage appearing across the insulator known as puncture voltage. This voltage causes breakdown and conduction of electricity inside the insulator. A puncture arc causes heat that can damage the insulator irreparably.
- **Flashover voltage**: Voltage that causes the air around the surface of the insulator to conduct electricity causes a "flashover" arc throughout the outside of the insulator. Usually these are designed to confront this without any damage.

In order to avoid damage, generally insulators are constructed with flashover voltage lower than puncture voltage. This will cause insulators to flash before they do damage.

**Table 3.7** Preventive maintenance of insulator

| Insulator maintenance schedule | | | |
|---|---|---|---|
| Inspection frequency | Items to be inspected | Inspection notes | Action required for unsatisfactory conditions |
| Quarterly | Cleaning of insulator | Either offline or online cleaning of insulator should be done. In a dusty environment it is essential to clean at short intervals | |
| Annually | Infrared scan | Scanning to be done at least once a year to detect the health of the insulators | Replace with new one |
| | Check health of the string | String health to be checked physically | Replace the portion not in compliance |

Flashovers and leakage currents can be caused by the presence of dirt, pollution, salt, and water in particular, which can create a conductive path across it. The magnitude of flashover voltage depends upon the presence of water and is reduced to more than 50% when the insulator is wet. In order to minimize the leakage current, the leakage path along the surface from one end to another is maximized and this path is also known as creep age length. These high-voltage insulators for outdoor purposes are designed to maximize creep age length. The surface is framed into a series of corrugations or concentric disk shapes so that the leakage path can be maximized. These generally contain one or more sheds which are nothing but cup-shaped surfaces facing downward, acting as umbrellas to ensure the surface leakage path under the cup also remains dry in wet weather. Safe or minimum creep age distances lie in the range 20–25 mm/KV and it should be increased in airborne sea-salt or high pollution areas. Please see Table 3.7.

## References


1. Power Sub-Transmission and Distribution systems-prevailing practices No. 198. New Delhi, December, pp 75–79
2. IS: 5613 (Part Iand II) Design installation and maintenance of lines up to and including 220 kV
3. Jones GR et al. (1993) Electrical engineer's reference book, 15th edn. Newness, Oxford, pp 22/1–22/18
4. Manual on Transmission Criteria, CEA (1985, May) p 15
5. Modern trends and practices in power sub-transmission and distribution systems, vol. 1. CBIP, Malcha Marg, Chanakyapuri, New Delhi
6. Brain F, Thornley J (1992) Lighting by design: a technical guide. Focal Press, Oxford
7. Wiring R, AS/NZS 3000:2000, Standards Australia, 2000, pp 132–143
8. Lerch E, Aggarwal RK (1995) Importance of system studies for industrial power system networks. In: Siemens Circuits, pp 12–19

9. Steward WE, Stubbs TA (1995) Modern wiring practice—design and installation. Newness, U.K.
10. Tripathy SC (1991) Electric energy utilization and conservation. Tata McGraw Hill, New Delhi, pp 73–105
11. IS: 10028, Code for practice for selection, installation and maintenance of transformers
12. IS: 5613, code of practice for design, installation and maintenance of over head power lines
13. IS: 6711-1972, Code for practice for maintenance of wooden poles for overhead power and telecommunication lines

# Chapter 4
# Best Practices in Operation and Maintenance of Distribution Substation Equipment and Auxiliaries

## 4.1 Distribution Substation

A power distribution substation is the heart of an electrical distribution network. Its prime objective is to step down power at the 66 or 33 kV level and distribute it in an 11 kV subdistribution network [1]. Normally these kinds of substations are located near load centers at the outskirts of cities. The maximum capacity of any 33/11 kV substation will be 60 MVA, either indoor or outdoor type, and also either air insulated (AIS) or gas insulated (GIS). These substations are constructed per the Indian standard, and other rules and regulations as per the latest amendment issues by the Central Electricity Authority of India (CEA). A 33/11 kV substation will have less than two transformers and at least two incoming feeders preferably from two different sources. Where both feeders are from the same source, each feeder will supply independent sections of the 33/11 kV substation, with these two sections isolated from each other by a bus coupler or isolator. All substations should have independent circuit breaker control of 33 kV incoming feeders, transformers, and 11 kV transformers. In general the following are the some of the important equipment/auxiliaries in a distribution substation.

1. Surge/lightning arrester
2. Instrument transformer (current transformer (CT) and potential transformer (PT))
3. Circuit breaker
4. Isolator and earth switch
5. Capacitor bank
6. Control and relay panels
7. Bus bar
8. Battery and battery charger
9. Earth grid and earthing system
10. Transformer



J. Raja et al., *Practices in Power System Management in India*,
Power Systems, https://doi.org/10.1007/978-981-10-2972-1_4

## 4.2 Surge/Lightning Arresters

**Description**

A lightning arrester (LA) protects substation equipment by discharging lightning and switching voltages over to earth. It consists of a series of spark gaps and several nonlinear resistances including thyrite and metrosil among others [2]. A nonlinear resister is one whose resistance is not constant but inversely proportional to the applied voltage and decreases rapidly as the voltage across it is increased [3]. The high surge voltage appears and allows the flow of heavy currents on the order of thousands of amperes, dissipates energy quickly and recovers again, and presents a high resistance value to the normal line voltage as soon as the surge has disappeared, so that any tendency of the arc to continue is immediately suppressed. In a system that has its neutral solidly earthed, the rated voltage of the arrester is usually taken as 80% of its maximum line-to-line voltage. In an unearthed system it is taken as 100% of line-to-line voltage because under fault conditions when one line is earthed, the arrester connected to the other two lines would be subjected to full line–line potential.

A 33 kV, 10 kA discharge current rating of LA should be installed for protection of switchgear, transformers, associated equipment, and 33 kV lines. Station class, heavy duty, gapless metal oxide (ZnO)-type surge diverters in general will be provided on the buses, high-voltage and voltage side of all transformers, and on the incoming terminations of 33 kV lines. The arresters will conform to IS 3070. The RMS voltage of LA will be 9 kV and the coefficient of earth not exceeding 80% as per IS: 4004, with all the transformer neutrals directly earthed.

1. **Types of Surge Arresters and Their Construction Details**

The development of surge arresters has taken place during from 1940 to the 1980s in the following stages.

1. 1940s—Surge arresters with rod gap
2. 1950s—Surge arresters with SiC discs
3. 1960s—Surge arresters with SiC discs and active gap
4. 1970s—Surge arresters with magnetically blown gaps for EHV-AC systems
5. 1975s—Surge arresters with rod gap and with metal oxide discs for EHV-AC systems.
6. 1979s—Gapless metal oxide arresters for HVDC
7. 1980s—Gapless SiC arresters with SF6 insulation
8. 1980s—Gapless ZnO arresters for various application

At present the following arresters are used.

1. Gapped silicon carbide surge arresters are used, also known as valve type or conventional gapped arresters.
2. Zinc oxide gapless arresters are used, also known as a ZnO arrester or metal oxide arrester. These are gapless and consist of a zinc oxide disc in series and have superior V/I characteristics and a high energy absorption level.

1. **LA Testing**

1. Routine test
2. Type test

   (i) **Routine test**

      (a) Sealing test
      (b) Measurement of reference voltage
      (c) Residual voltage test
      (d) Internal insolation test

   (ii) **Type test**

      (a) Insulation withstand test on the arrester housing
      (b) Residual voltage test
      (c) Long duration current impulse test
      (d) Operation duty test
      (e) Short-circuit test
      (f) Polluted housing test
      (g) Internal partial discharge test
      (h) Bending moment
      (i) Environmental test
      (j) Seal leak test
      (k) Radio inference test

1. **Maintenance of Surge Arresters**

   The following are the usual causes for the failure of a surge arrester.

1. POWERGRID case histories show about 40 LAs failed due to moisture entry.
2. Approximately 50 LAs have been removed based on third harmonic resistive current measurements. During investigations, moisture entry was found to be the main reason for THDC violations.
3. Mostly during the switching operation 80 LAs have failed.
4. During PIR removal field studies, LAs have not conducted even on a single occasion in worst conditions (1.95–2.05 PU) indicating switching surge current less than 70–80 A.
5. During failure investigation with manufacturers in 2003/2004, most LA failures have occurred due to moisture entry and not due to ageing/conduction.
6. For transformer/reactor LAs, TOV requirements are fewer compared to line LAs and hence rating may be different for both LAs. Moreover, in most cases, for transformers only the residual surge travels to the transformer.

1. **Causes and Nature of Failure**

The surge/lightening arrester responds to overvoltages without any time delay. According to system requirements the rated voltage, continuous operating voltage, energy handling capability, nominal discharge current, and other characteristics of the LA arrester will be chosen [4, 5]. Normally LA rarely fails, however, it has been designed for a particular number of operations. The same has to be properly checked during selection of a LA. The number of the discharge cycle counter should be checked on a daily basis and when the counter is in the "RED" zone it should have changed with immediate effect. If the surge arrester is being used at abnormal service conditions such as excessive deposits of smoke, dirt, salt spray, or other conducting material, the surge can be bypassed and the arrester stop failure of the equipment.

2. **Primitive Maintenance of LA**

In general primitive maintenance of a lightning arrester doesn't occur. Once it fails it has to be replaced. However, during puncture the porcelain insulator needs to be replaced. The failure of nonlinear resistance in LA will have to be fixed. Hence in the case of this failure, the LA has to be replaced as a whole. It is important to choose the appropriate class and rating of the LA during selection, especially considering environmental conditions. Please see Table 4.1.

**Table 4.1** Preventive maintenance schedule of LAs/surge arresters

| Inspection frequency | Items to be inspected | Inspection notes | Action required for unsatisfactory conditions |
|---|---|---|---|
| Weekly | Reading of leakage current | The reading to be noted on weekly basis. However, after heavy lightning it should be checked. If current shoots in red zone, then that particular LA is to be replaced as early as possible | – |
| | Visual inspection | Visual inspection of insulator to be made at least once a week. In case of any physical damage, it has to be replaced | Replacement of LA |
| Monthly | Connection tightness | The tightness of each individual connection has to be assured on a monthly basis | Necessary nuts, bolts, and washer may be replaced in the case of proper connectivity |
| | Checking IR values | The LA earth connectivity has to be checked. Resistivity between stack to stack and between each stack to earth by suitable Megger | Failure to get the desired value will cause check in connectivity, earth pit, and LA contact |
| | Surge counter test | Apply 230 V AC supply across the counter and check pointer movement in clockwise direction | Replacement of counter |
| Half-yearly | Insulator cleaning | The insulators should be cleaned with a clean smooth cloth. Care should be taken so that no scratch appears over the insulation | Cleaning may be done using detergent or replacement may be done |
| Yearly | Measurement of reference voltage (V-ref) | The reference voltage is the sum of all the individual reference voltages. The peak voltage is independent of power frequency voltage, and the current of the arrester is divided by $\sqrt{2}$ measured at the reference current | – |

(continued)

**Table 4.1** (continued)

| Inspection frequency | Items to be inspected | Inspection notes | Action required for unsatisfactory conditions |
|---|---|---|---|
| | Power and frequency reference voltage measurement | On the complete arrester at the reference current measured at the bottom of the arrester. The measured value will be within the range specified by the manufacturer. For multiunit arresters the value may deviate from the reference voltage of the arrester | – |
| | Partial discharge test | For the partial discharge test the power frequency voltage applied to the complete arrester, along with its end fittings, will be increased up to its rated voltage and within 10 s decreased to 1.05 times its continuous operating voltage. At that voltage the partial discharge level according to IS 6209 will be measured | – |

3. **Predictive Maintenance**

After collecting performance records of the surge diverter on a regular basis, when it is found that the counter goes to the red zone it is necessary to replace the arrester at the earliest.

## 4.3 Instrument Transformer (CT and PT)

**Description**: The instrument transformer is used to step down high voltage or current from a transmission/distribution system into a low value for metering and protection purposes. These transformers reduce the high current or high voltage connected to their primary windings to the standard low values in the secondary. They are expected to be maintenance-free during their service life. They are of minimum oil type and hermetically sealed and from the application point of view, they are divided into three main categories.

**Metering Type**

The specified performance of CT is to be maintained in the range normally 5–120% of the rated current. The CT cores should be such that they saturate at the instrument security factor (ISF) for safeguarding the instrument from getting damaged under fault conditions [6, 7]. The PT designed for metering is required to perform as specified within the voltage range near the normal rated voltage of 80–120%.

**Protection Type**

The main performance requirement of protection class CTs is that its cores should not get saturated below its accuracy limiting factor (ALF) up to which the primary current should be faithfully transformed to the secondary, maintaining the specified accuracy. During fault conditions, the CT primary carries a very high current and first few cycle have DC component, which may saturate the core. Behavior of the cores in such condition should be such as to avoid becoming magnetized and to return to normalcy (demagnetized stage) soon after clearing the fault.

**Load Survey Type**

This is a mixture of the above two categories and is usually used for economic management of industrial loads. The output of this CT is connected with various distribution management systems.

## *4.3.1 Outdoor Type Instrument Transformer*

These are normally used in power substations (Fig. 4.1). While designing, the following factors are taken into consideration.

1. Effect of atmosphere environment.
2. Insulation to withstand network fluctuation.
3. To avoid ingression of moisture these are oil filled.

(a) Current Transformer(CT) (b) Potential Transformer (PT)

**Fig. 4.1** Current transformer and potential transformer

### 4.3.2 CT Type

1. **Window CT**: It has no primary winding construction and is installed near the primary conductor.
2. **Bushing CT**: Window CT is not accessible because it is built near bushing.
3. **Bar CT**: A window CT but has a permanent bar installed as a primary conductor.
4. **Wound CT**: Like a normal transformer, the wound CT features primary and secondary winding. This CT is mostly used for lower conversion ratio of current and is generally applicable in matching various CT summing ratios to compensate lower current in CT secondary circuits or to separate similar CT circuits. The instrument transformer is supposed to be maintenance-free and hence there is no scope of filtering or change of oil during its life. This makes it essential to hermetically seal the transformer to avoid breathing atmospheric air.
5. **Accuracy Limit Factor (ALF)**: It is the ratio of the largest current value to CT rated current up to which the CT must retain the specified accuracy. Example: –CT –5P20, 5 VA, ALF = 20.
6. **CT Core Identification per Class**:
    1. Class—0.2 s, 0.5 s, and 1.0 s: Metering core
    2. Class—5P10, 5P20, etc.: Backup protection core (O/C and E/F protection)
    3. Class—PS: Primary protection core (differential, distance, REF, etc.)

1. **Causes and Nature of Failure**

In general all this equipment is maintenance-free throughout its lifespan. However, this instrument fails due to the following factors. Some of these devices are outdoor types and are designed to withstand environmental conditions. Porcelain insulators are being used for external isolation between live and ground. They provide an outer casing for all atmospheric conditions including rain, dust, chemical contamination, wind, sun, and so on. However, loose connections between the insulator and tank, breakage of washers between them, and oil leakages from the tank are common. It is necessary to maintain this equipment periodically.

In a power system a frequent number of power surges are generated due to switching, sparking, and other network disturbances in the system. These environmental causes make core saturation which ultimately damages the equipment by disturbing measurement accuracy [8]. Due to these power surges the insulation between the primary and secondary circuits becomes damaged and ultimately damages the equipment.

2. **Primitive Maintenance of Instrument Transformers**

For a smooth and reliable power system it is important to keep all the equipment healthy throughout the lifecycle. Hence as soon as the equipment fails it has to be replaced. However, Table 4.2 shows the measures that should be taken for preventing failure of this equipment.

**Table 4.2** Preventive maintenance schedule of instrumentation transformer

| Inspection frequency | Items to be inspected | Inspection notes | Action required for unsatisfactory conditions |
|---|---|---|---|
| Weekly | Physical verification and cleaning of bushing | To be carried out once a week for checking secondary connection, oil leakage, crack in porcelain, and so on. The bushing has to be cleaned | Corrective action should be taken immediately |
| Monthly | Oil level and oil leakage | For outdoor type CT/PT units oil level and oil leakage have to be checked | Refill with transformer oil. For oil leakage immediately check or replace the unit |
| | Power connection tightness | All connecting points to be checked for tightness | Replace nuts, bolts, or washers as required for proper tightness |
| | Proper earthing of body connection and CT secondary core star point | Proper earthing of body to be checked and maintained | – |
| Yearly | Insulation resistance checking | All the insulation resistance (i.e., resistance between primary and secondary, as well with earth has to be measured) | Wherever required, change insulation or regain insulation level |
| | Sealing with metallic bellows | The movement of metallic bellow has to be checked | – |
| | Sealing with nitrogen cushion | The pressure for nitrogen gas and sealing to be checked | – |

3. **Preventive Maintenance of Instrument Transformer**
4. **CT Testing**
1. **IR Testing**:

    (a) Primary to earth by 5 kV Megger
    (b) Secondary each core to earth by 500 V Megger
    (c) Primary to secondary by 5 kV Megger
    (d) Secondary core to core by 500 V Megger

2 **Polarity Test**: For carrying out this test, we require one 1.5 V cell, DC analog ammeter. This ammeter has to be connected across S1 and S2 and 1.5 V DC is given across the terminal. By making the above connection, if there is positive deflection of the ammeter, then polarity is confirmed.

3 **Ratio Test**: Inject current in primary winding and measure the induced secondary current for different current readings and verify with CT ratio.

4 **Knee Point Check for PS Class Core**: Inject 230 V variable AC voltages in secondary core with ammeter in series. At a certain stage, with 10% increase in voltage, the current shoots up almost 50% and it is called knee point voltage.

After carrying out this test the CT is demagnetized by the gradual reduction of voltage to zero.

5 **Winding Resistance Test**: Measure secondary winding resistance by micro ohm meter.

### 4.3.3 *PT Testing*

1. **IR Testing**:
   (a) Primary to earth by 5 kV Megger
   (b) Secondary each core to earth by 500 V Megger
   (c) Primary to secondary by 5 kV Megger
   (d) Secondary core to core by 500 V Megger
2. **Ratio Test**: Inject AC variable voltage in primary winding and measure induced secondary voltage at different voltages and verify the same with PTR.

3 **Predictive Maintenance**: Instrument transformers are used for measurement and protection purposes. Accuracy of operation of this equipment is very essential. In order to operate with high accuracy predictive maintenance is necessary. If during testing and calibration it is found that the equipment is going to cross the accuracy level in the near future due to ageing, saturation, decay in winding, and so on, immediate measures should be taken in order to avoid disturbances and have reliable operation.

1. **Circuit Breaker**

**Description**: The circuit breaker is used to close or isolate the circuit in normal and abnormal conditions and to protect the electrical equipment against the fault.

Circuit breaker classifications follow.

## 4.4 Classification of Circuit Breaker

Circuit breakers can be classified according to their arc quenching medium, application, and use in sites. Based on the use of the breaker it can be classified as follows.

1. **Outdoor-type breaker**: In breakers the earth quenching mechanism is present at the switch yard. Only the circuit breaker control mechanism is present inside the control room.
2. **Indoor-type breaker**: In these varieties the earth quenching mechanism is placed inside the control room. HV circuits are routed to the circuit breaker operating chamber inside the control room.

However, based on different arc quenching mechanisms, circuit breakers are classified as follows.

1. **Bulk oil circuit breaker**: Contacts are separated inside a steel tank filled with transformer oil used for arc quenching.
2. **Minimum oil circuit breaker**: Contacts are separated in an insulated housing (interrupter) filled with transformer oil used for arc quenching.
3. **Air blast circuit breaker**: It utilizes high-pressure compressed air for arc extinction.
4. **SF6 gas circuit breaker**: Sulphur–hexafluoride gas is used for arc extinction in this breaker. The SF6 breaker has an advantage that the rate of restricting voltage is zero and hence the burning of male/female contacts is less [9]. The operating mechanism is of two types:

    1. Movement of contacts is controlled by a spring mechanism (spring operated).
    2. Movement of contacts is controlled by air pressure (pneumatic operated).

5. **Vacuum Circuit Breaker**: In this breaker, the contacts are housed inside a permanently sealed vacuum interrupter. The arc is quenched as the contacts are separated in high vacuum.

1. **Causes and Nature of Failure**

During normal operating conditions, circuit breakers can be opened or closed by an operator for the purpose of switching and maintenance. During abnormal or faulty conditions the relay senses the faults and closes the trip circuit of the breaker; thereafter the circuit breaker opens the circuit. As the relay contacts close the trip circuit, the operating mechanism of the circuit breaker opens the contacts and an arc is drawn between them. The arc is extinguished at some natural current zero of the AC wave. The process of current interruption is complete when the arc is extinguished and the current reaches final zero value.

However, in the process carbon is deposited over the arcing contacts, main contacts, and auxiliary interlocking contacts. This ultimately causes delay in fault clearing time as delay in execution of the trip command.

The contacts are burnt/worn out or incur excessive roughness due to regular use. In these conditions contacts need to be replaced with new ones. With regular use, the spring becomes staggered due to continuous compression or expansion causing delay in the operating mechanism. This needs to be checked and updated. Necessary lubrication at all moving contacts should be used for smooth nonsluggish movement.

In the circuit breaker a numbers of nuts, bolts, and washers are used. Due to high accelerated movement with huge force, these washers and nuts may loosen, causing displacement of the equipment from its original trajectory. In order to have smooth and accurate movement these need to be tightened. The circuit breaker carries a huge discharging current. For safe operation and long life of the equipment earth connections need to be checked at regular intervals.

## 4.5 Primitive Maintenance of Circuit Breaker

Oil, SF6 gas, and air blast circuits require a specified quantity and pressure of quenching medium for reliable and smooth operation. Quantity and pressure should be maintained at regular intervals [10]. Indicator monitoring of these parameters should be recorded on a regular basis. Breakdown or bends in the circuit breaker contacts need to be attained in immediate effect. The contact force needs to be measured in the case of the vacuum circuit breaker on a regular basis and necessary correction taken.

1. **Preventive Maintenance of Circuit Breaker**

The circuit breaker is the most important protective device on the power distribution network. Its preventive maintenance is to be carried out at highest priority (Table 4.3).

**Table 4.3** Preventive maintenance schedule of circuit breakers

| Inspection frequency | Items to be inspected | Inspection notes | Action required for unsatisfactory conditions |
|---|---|---|---|
| Monthly | Tightness of control wire, power connection, nuts, bolts, and washer | Tightness needs to be checked on a monthly basis and corrected | Replace wherever required |
| | Gas pressure check | In SF6 types of breaker gas pressure needs to be checked at regular intervals | Leakage needs to be filled in |
| | Air pressure check | Air pressure needs to be checked in pneumatic operated breaker | Leakage needs to be filled in |
| Half-yearly | Mechanical operation | Mechanical operation needs to be checked by making false trip command. Duration of operation needs to be recorded | Fault to be analyzed and rectified instantly |
| | Insulation resistance | Checking of insulation resistance of each pole and to ground across terminals | – |
| | Cleaning and lubrication | Cleaning and lubrication of contract parts | Replace the burnt-in/worn out contacts |
| Yearly | Dielectric strength of oil | In oil circuit breaker this has to be carried out at least on a yearly basis | Replace with new transformer oil |
| | Porcelain insulator | Cleaning and checking for the isolators | Replace if any failure found |
| | Overhauling of breaker | Tightness of nuts, bolts, and washer needs to be checked and done | – |

(continued)

**Table 4.3** (continued)

| Inspection frequency | Items to be inspected | Inspection notes | Action required for unsatisfactory conditions |
|---|---|---|---|
| | Checking pole discrepancy | Checking controls, interlocks, and protections such as checking pole discrepancy system (i.e., whether all three poles are getting ON–OFF at the same time) | Necessary correction in setting to be done |
| | Operating time | The circuit breaker needs to be checked for the time required for actual close | Necessary correction in relay setting to be done |

2. **Predictive Maintenance**

Each circuit breaker has a counter recording the number of operations done with the breaker. In the manufacturer's user manual it is clearly mentioned that after a certain number of operations the arc quenching medium needs to be replaced. The necessary guidelines given in the manufacturer's user manual are to be followed strictly. The delay in time from the fault recording and circuit break needs to be checked properly. For higher delay time relay settings and operating time need to be checked and corrected.

## 4.6 Isolators and Earth Switch

### *4.6.1 Classification of Isolators*

The isolator (Fig. 4.2) is the device that makes and breaks circuits in no-load condition. The isolators provided in the substation are for disconnecting the line and equipment from the bus bar or from incoming/outgoing lines [11]. These are off-load switches and not used as load break switches.

The various types of isolators are as follows.

(a) **Center Break Rotating Type Isolator**: This type of isolator is used for bus bars and incoming lines. In this type both side insulators rotate and the break is achieved at the center.

(b) **Double Break Rotating Type Isolator**: This type of isolator is used for connecting or disconnecting the bus bar and equipment or connecting or disconnecting the line from the substation. In this type, the center insulator rotates and the copper tube or blade is the moving contact. The other two side insulators are fixed and they carry fixed contacts. The center insulator rotates affording double break operation on opening the switch.

**Fig. 4.2** Isolator

(c) **Pantograph Type Isolator**: In this type of isolator a break is vertical. The isolator touches a bus bar when it is in the ON position and it comes down vertically when it is OFF, disconnecting the bus bar connection. These types of isolators are in use in our system at various 400 kV substations.

(d) **Earthing Isolators**: The earthing isolators are required to earth the line when it is switched OFF, to discharge the line to earth capacitive voltage. This is very important when people are working on the lines for maintenance purposes. The earthing blades are attached to incoming or outgoing feeder isolators and they can be made ON only when the line isolator is OFF. Such type of interlocks is very necessary to avoid earthing of a live line.

1. **Causes and Nature of Failure**

Isolators are off-load switches and in normal operation heavy current flows through them. However, a small gap in the contacts of the isolator causes heavy sparks and burning of isolator contacts.

In general the male/female contact of the isolator develops erosion and burnout ultimately causing heavy sparks and breaking of the isolator moving contact. As the handle, rotating rod, and moving mechanism are all outdoor mechanical devices they decay with the age of the isolator [12, 13]. This will cause loosening in the contacts, nuts, bolts, and the like.

2. **Primitive Maintenance of Isolators**

Isolators can be maintained after reporting a fault in any of their parts. It is essential to keep the contacts healthy and proper at all times. After establishing a burning mark or corrosion on an isolator contact it is essential to clean the contacts on a regular basis or change the corroded contacts for smoother operations.

During isolator operation (manual as well as motorized) the moving part deteriorates causing loosening in contacts, improper alignment, and so on. This can be corrected on a regular basis by tightening nuts, bolts, and washers. Lubrication of all the moving parts on a regular basis will also reduce the deterioration effect.

A preventive maintenance schedule for isolators is shown in Table 4.4.

3. **Predictive Maintenance**

After getting a poor result during Meggering of earth resistance from the isolator, it is essential to check for the loose path of earth. All the nuts, bolts, and connectors need to be corrected and healthy. In the process of regular scheduled maintenance whenever improper contacts are found it is essential to replace them with a new matching component. This will ultimately nullify the gap during the closed condition and a spark will be isolated.

**Table 4.4** Preventive maintenance schedule of isolators

| Inspection frequency | Items to be inspected | Inspection notes | Action required for unsatisfactory conditions |
|---|---|---|---|
| Weekly | External particle over isolator | Cobwebs, bird's nests, paper kites, loose threads surrounding the insulators to be removed and cleaned at the earliest | – |
| | Crack in insulator | Cracks or broken insulator on isolator base to be replaced immediately | Immediate replacement |
| Monthly | Checking of the male/female contacts | Male/female contacts should be checked once a month. The burn marks, rust, and other external particles removed | Replacement of contact |
| | Lubrication | All moving parts should be lubricated on a regular basis for smooth movement | – |
| | Panel indicator and control wiring | The panel indicator (i.e., semaphores and bulbs, etc.) should be checked at regular intervals. The wiring for motorized isolator also has to be monitored | – |
| Half-yearly | Alignment of moving contact | The alignment of the male/female contact should be checked and corrected | – |
| | Cleaning of insulator | Insulators should be cleaned with a clean and smooth cloth | Replacement of insulator |
| Yearly | Tightening of earthing connection | Earthing connection should be checked at least once a year by Meggering the resistance level. The same should be recorded and preventive steps taken | – |
| | Interlock check | Isolators are normally interlocked with earth switch and circuit breaker. This has to be checked for clear operating mechanism | – |

# 4.7 Capacitor Bank

## *4.7.1 Description*

In any power utility, maintaining a stable power supply at the proper voltage is always a problem. Due to lot of inductive load, the reactive power flow takes place in the system, resulting in the lowering of system voltage and an increase in transmission and distribution losses [14]. The HT capacitor provides an interim solution in improving power system stability, voltage, and the power factor. The HT capacitor bank also compensates the losses occurring in the transmission lines. The capacitor unit has one steel container, two bushings, and several capacitor elements enclosed in the unit. A single HV capacitor may have a capacitance of 5 KVAr–200 KVAr. Several identical units are mounted on insulator racks and connected in series parallel combination to obtain a high-voltage capacitor bank (Fig. 4.3).

Before commissioning, Megger the capacitor bank between phases and earth. The Megger reading for an individual capacitor should not be less than 50 MΩ. For more than one unit in parallel, the minimum acceptable Megger value can be derived by dividing 50 MΩ by the number of units connected in parallel. Before switching on the capacitor, bus voltage, system incoming load current, and power factor can be noted. After energizing, check that the capacitor draws almost a balanced current in all three phases and is near its rated value. Note the change in bus voltage, load current, and system power factor. Normally after capacitors are energized, there will be a little rise in bus voltage and some reduction in system load current and improvement in power factor. In the case where load current increases instead of reducing, it shows that the connected capacitors are more than

**Fig. 4.3** Capacitor bank

required for the load and in this case the power factor will be leading. When the residual voltage factor (RVT) is used for imbalance protection, measure open delta voltage, which should be negligible. In the case where capacitors are connected in double star with neutral CT, the current on the secondary side of neutral CT can be measured, which should also be negligible.

1. **Causes and Nature of Failure**

A fuse is provided for each capacitor in the bank. The fuses will be external type for an 11 kV capacitor bank. Sufficient clearance is maintained between bodies to the line terminal by the proper choice of capacitor unit with an external fuse used to avoid bird faults. When the voltage level in the capacitor bank is more than 11 kV it is provided with an internal fuse type. In the case of fault, the faulty element will automatically go out of circuit.

Discharge resisters are provided within the capacitor unit to ensure safety after de-energizing of the capacitor (within or less than 5 min, the rated voltage of 50 V is achieved by decreasing the residual voltage from crest value). The power loss in these resisters is negligible. Each capacitor bank is protected against lightning by a gapless zinc oxide arrester. The capacitor protection equipment includes overcurrent, earth leakage, and protection to detect imbalance loading due to abnormal conditions.

2. **Primitive Maintenance of Capacitor Bank**

A capacitor bank is normally provided at the 11 kV side of a 33/11 kV substation. Due to the internal chemical effect the bushing/leads of the capacitor bank erode. An erosive lead causes poor conductivity and ultimately breaks the flow. In this regard its bushing and leads need to be checked and cleaned at a regular intervals for long use. The capacitor bank consists of several units of small capacitors connected in series/shunt. Each unit should contain its separate fuse set either internally or externally. During fault conditions the fuse blows out and isolates the capacitor from faulty conditions. However, after restoration of the fault before going for normal use these fuses have to be restored for smooth operation. On the eve of failure of an individual capacitor unit it has to be replaced with a healthy unit before the whole bank is put in operation. A preventive maintenance schedule is given in Table 4.5.

3. **Predictive Maintenance**

Whenever the capacitance of the whole bank shows unexpected results, the exact failure of the device is to be identified and should be replaced and the final result restored.

**Table 4.5** Preventive maintenance schedule of capacitor bank

| Inspection frequency | Items to be inspected | Inspection notes | Action required for unsatisfactory conditions |
|---|---|---|---|
| Weekly | Cleaning of bushing/leads | The terminals need to be cleaned at regular intervals at least once a week | If not cleanable, terminals should be changed |
| | Check for leakage | Capacitor needs to be checked for any kind of leakage | – |
| | Tightness of all connections and earth connection | These need to be checked at regular intervals. Earth connection needs to be checked. However, all equipment should connect as per the schema | Correct as soon as identified |
| Quarterly | Charging discharge cycle | It is important to complete capacitor bank discharge once a quarter. During discharging never short the terminals. It has to be discharged with proper procedure | – |
| | Checking value of capacitance and discharge resister | It needs to be checked at regular intervals to determine its health | Replace if fails to give the value |
| | Oil break down voltage | Oil BDV needs to be checked for series reactor and NCT/RVT | – |
| Yearly | Protection devices | Connection with all protection devices needs to be verified to avoid any damage in abnormal conditions | – |

## 4.8 Control and Relay Panels

### *4.8.1 Classification of Control and Relay Panels*

Control or relay boards are built up by using the requisite number of self-contained sheet steel cubicles, comprising a front panel to carry the control apparatus. Depending upon the size of the substation the control and relay board may incorporate the following.

1. Simplex type: The hinged or removable back cover gives access to interior wiring and cable termination.
2. Duplex panels: Panels are arranged back to back in corridor formation, and a door is then fitted at each end.

A mimic diagram representing main circuit connections is incorporated on the front panel. It is a single-line diagram incorporated on the front side of the control panel [15]. This diagram represents the actual physical position of various HT electrical equipment in the substation yard along with status of equipment and ON and OFF positions of various breakers and isolators through semaphore or lamp indication. The circuit breaker control switch is fitted on front. Normally the switch is in normal (center) position. The handle is moved to the right or left to initiate close or trip operations.

Indication lamps mounted for various purposes follow a standard color code.

Red: CB or switch CLOSED
Green: CB or switch OPEN
White: Trip circuit healthy
Amber: Alarm indication (i.e., CBs tripped on fault).

### *4.8.2 Colors for Internal Wiring*

1. **Red**: Phase connection, either directly connected to the primary or secondary circuit of CT and PT
2. **Yellow**: Phase connection, either directly connected to primary or secondary circuit of CT and PT
3. **Blue**: Phase connection, either directly connected to the primary or secondary circuit of CT and PT
4. **Black**: AC neutral connection, star point connections of secondary circuit of CT and PT, and connections in AC and DC circuit
5. **Green**: Connections to earth
6. **Grey**: Connections in DC circuit

Each wire should have a letter to denote its function. DC supply from +ve source should bear an odd number and from −ve source should bear an even number.

**CT Secondary Terminal**: S2 of all protection and metering cores are shorted in the CT junction box. Only one common S2 wire along with S1 wires of all three phases of CTs are brought to the control relay panel. Earthing of S2 wires is done at one end (preferably at the CRP end).

**Common Ferrule Numbers Used in Wiring**

1. CT secondary connection for primary protection such as differential, distance, REF relay). Small "a" is used for PT secondary connection in PT terminal box.
2. Bus bar protection (CT secondary connection). B for B phase indication.
3. Backup protection (CT secondary connection for O/C and E/F relay).
4. Metering (CT secondary connection).
5. Metering and protection (PT secondary connection).
6. AC connection.
7. DC connection (before fuse).
8. DC connection for control (after fuse).
9. DC connection for indication (after fuse).
10. Motor supply (spring charging motor in circuit breaker).
11. RTCC (tap changer) connection. Also for denoting AC neutral connection.

12. PT primary connection and DC circuit of bus bar protection scheme.
13. R phase indication.
14. CT secondary connection in terminal box.
15. Circuit breaker auxiliary contacts.
16. TB numbering.
17. Y phase indication.

1. **Causes and Nature of Failure**

Various controlling and protective devices are placed in these panels. Physically large numbers of connecting wires are linked with the control panel. It is important to have proper identification of each control cable with the connected devices. Hence ferruling plays an important role in failure cause analysis of the control panel. However, the control panel seldom fails; the devices connected with the control and relay panel fail. The power supply to the panel is normally given from the adjunct DC supply system [16]. Control and relay equipment sometimes fail due to failure of a quality power supply. Some types of faults are shown in Table 4.6.

**Table 4.6** Types of identified faults

| S. No. | Types of identified faults | Works to be attended |
|---|---|---|
| A | CT Circuit | |
| 1 | Noise in CT | Open circuit of CT secondary circuit |
| | | Loose connection in CT secondary circuit |
| | | Ammeter switch fault causing CT secondary open circuit |
| | | Loose stampings of CT |
| | | In case CT is provided with a surge diverter in primary winding, this might have been disconnected |
| 2 | Ammeter not recording properly | Ammeter may be faulty |
| B | PT circuit | |
| 1 | Voltmeter not showing properly | Check fuses on HV and LV sides of PT |
| | | If above is OK, voltmeter may be faulty |
| | | Loose connection in PT circuit |
| 2 | Energy meter recording less | Same as above |
| 3 | Energy meters not recording properly | If beyond adjustment, energy meter may be faulty |
| C | DC protection circuit | |
| 1 | Nonworking of trip healthy indication | Fuse of bulb or bulb may be loose |
| | | Loose connection |
| | | Resistance may be open circuit |
| | | Alignment of auxiliary contacts of circuit breaker disturbed |
| | | Trip coil open |
| | | DC fuse may be loose or blown off |
| | | DC link may be loosely fitted |
| | | Pushbutton may not be making good contact after pressing |

(continued)

**Table 4.6** (continued)

| S. No. | Types of identified faults | Works to be attended |
|---|---|---|
| 2 | Nontripping of breaker | Either loose fitting of fuse and link or blowing of control fuse |
| | | Loose connection |
| | | Open circuit of trip coil or trip coil might have been burnt |
| | | Alignment of auxiliary contacts of circuit breaker disturbed |
| | | No free movement of plunger in trip coil assembly |
| | | Mechanical trouble in breaker |
| | | The local/remote switch may be in local position while tripping breaker from remote |
| | | Trip switch or pushbutton may be faulty |
| 3 | Nonclosing of breaker | Either loose fitting of fuse link or blowing of control fuse |
| | | Loose connection |
| | | Open circuit of closing coil or closing coil might have been burnt |
| | | Alignment of auxiliary contacts of circuit breaker disturbed |
| | | No free movement of plunger in closing coil assembly |
| | | Spring might have been charged |
| | | Nonresetting of trip/master relay |
| | | Nonclosing from remote end, if local/remote switch may be on local position |
| | | Closing switch or pushbutton may be faulty |
| | | In case of capacitor breaker, auxiliary relay in the timer circuit might not have reset as per the time setting given to timer |
| 4 | Tripping of breaker without indication | Due to short in DC positive |
| | | Leakage of DC |
| | | Due to low pressure |
| 5 | Closing of air blast circuit breaker without indication | Due to short in DC positive |
| | | Leakage of DC |
| | | Due to low pressure |
| 6 | Spring charging motor doesn't start | Either loose fitting of fuse and link or blowing of fuse |
| | | Failure of AC supply |
| | | Failure of limit switch CB |
| | | Defective motor |
| 7 | Flashover and damage of winding | Failure of insulation |
| | | Short circuit due to lizards/rats, etc. |
| D | Annunciation circuit | |
| 1 | Nonworking of bell | Alarm DC fuse and link might have been fitted loosely or fuse might have blown off |

(continued)

**Table 4.6** (continued)

| S. No. | Types of identified faults | Works to be attended |
|---|---|---|
| | | Loose connection |
| | | Disturbing of bell adjustment |
| | | Burning of bell coil |
| | | Auxiliary relay provided for bell circuit may not be working properly |
| | | Sealing (hold on) supply getting to the aug. relay through "accept" pushbutton might have disconnected due to faulty accept pushbutton |
| 2 | Continuous ringing of bell | DC leakage |
| | | Adjustment of auxiliary relay contact is disturbed |
| 3 | Nonresetting of bell | Accept pushbutton faulty |
| | | DC leakage |
| | | Auxiliary relay faulty |
| 4 | Flasher working on without fault | DC leakage |
| 5 | Nonresetting of indication | DC leakage |
| | | Reset pushbutton faulty |
| E | Indication circuit | |
| 1 | Indication lamp not indicating for breaker ON/OFF position | Loose fitting of fuse and link provided for indicating circuit or this fuse might have blown off |
| | | Lamp may be loosely fitted or fused |
| | | Loose connection |
| | | Auxiliary switch may be defective |
| 2 | Semaphore not working | Loose fitting of fuse and link provided for indicating circuit or this fuse might have blown off |
| | | Loose connection |
| | | Defect in auxiliary switch of breaker or isolator |
| | | Semaphore coil burnt |
| | | Nonavailability of auxiliary contact of isolator |
| F | Relays | |
| 1 | Bad operation of relay | Relay defective, needs calibration |
| 2 | Flag not resetting | Mechanical defect in flag mechanism |

## 4.9 Bus Bar

In an electric power system, bus bars are considered as the source of electric power. They are composed of aluminum or copper conductor material, where all the loads are interconnected via insulators. The bus bar is mainly deployed in the electrical substation to tie the incoming and outgoing lines of the transformer and transmission system; it also connects the generator and transformer in power-producing stations [17]. For industrial applications, aluminum smelter-type bus bars are used

**Fig. 4.4** Bus bar arrangement

to carry large amounts of current for their electrolytic processes. Bus bar size is an important parameter that determines the capacity of the maximum amount of current flow. The cross-sectional area of the bus bar varies depending on the application; for example, a 10 mm$^2$ cross-sectional bus bar is used for small current flow applications, whereas a 1000 mm$^2$ or more cross-sectional bus bar is used for electrical substations.

A bus bar structure may be completely surrounded by insulation or supported by insulation. The bus bar shown in Fig. 4.4 is sheltered by a metal enclosure or by elevation out and a neutral bus bar is also insulated to avoid accidental contact. Earth bus bars are directly bolted into metal chassis for their protection. The bus bar structure may be covered in the busducts, segregated phase busducts, isolated phase busducts, and so on.

Clamps or bolts are used to connect bus bars with other electrical appliances. A silver-plated surface is done in high current bus joints to minimize contact resistance. For extra-high voltage above 300 kV, connection fittings are intended to reduce corona which may be a source of electromagnetic interference with nearby connections. The major types of bus bars are shown in Table 4.7 and a preventive maintenance schedule in Table 4.8.

The rigid bus bar has a porcelain insulator, the strain bus bar is flexible, and a stranded conductor is used to connect between the substation metal frame structures by suspension type insulators. Generally bus bars consist of aluminum or copper materials. The insulated phase bus bar is surrounded by a grounded metal shield and supported with an insulator; this approach helps in eliminating short-circuits among adjacent phases. A sulphur hexafluoride-insulated bus bar is a big metal tube filled with high-pressure sulphur hexafluoride gas.

**Table 4.7** Major types of bus bars

| S. No. | Type of bus bar | Usage |
|---|---|---|
| 1 | Rigid bus bars | Low, medium, and high voltage |
| 2 | Strain bus bars | Mainly for high voltage |
| 3 | Insulated-phase bus bars | Medium voltage |
| 4 | Sulphur hexafluoride (SF6)-insulated bus bars | Medium- and high-voltage systems |

**Table 4.8** Preventive maintenance schedule of bus bar

| Inspection frequency | Items to be inspected | Inspection notes | Action required for unsatisfactory conditions |
|---|---|---|---|
| Weekly | Physical Inspection | Visual inspection of health. All the joints are proper collected | – |
| | Burnt mark, color of bus bar | During visual inspection joints on the bus bar to be checked | Spark may be checked on bus bar |
| Quarterly | Earthing measurement | Meggering to be done for measuring the connectivity between devices | – |
| | Inspection of joints | Inspection of all jointing equipment and check joints | – |

## 4.10 Battery and Battery Charger

### *4.10.1 Classification of Battery Types*

The battery is used to transform chemical energy into electrical energy by electrochemical reaction. The cell is the basic component in the battery, where cells are connected in series or parallel to form the battery unit.

The battery and battery charger are small but important pieces of equipment in electrical substations (Fig. 4.5). In lieu of a more reliable power supply source, a DC supply is used in place of the normal AC supply. If there is a blackout, AC power will not be available and the electrical protection devices will not operate. To avoid any such situation, a DC supply is used [18]. The DC supply is provided by the batteries and the charger is used to keep the batteries healthy by charging the batteries in float/boost mode as required.

In a substation, the battery is commonly used to power the supply relay and breaker tripping mechanism. These monitoring device operations are ensured by a trustworthy battery power supply, leading to successful operation of that device. Thus it is essential to maintain a reliable battery power supply by proper inspection for proper operation of the monitoring device. In general terms, the capacity of a cell/battery is the amount of charge available expressed in amp-hours.

**Fig. 4.5** Battery and battery charger

Battery cell ability is estimated based on the discharge of the battery at a constant current rate and is multiplied by the time required to reach its terminal voltage of 1.75 V. In a substation 33/11 and 66/11 kV rating system, the standard battery voltage rating of 30, 110, or 220 volts is deployed for storage usage. Those batteries usually have a rating of 45 Ah (min), 24 V to meet the load requirement of the substation. Table 4.9 gives the cell voltage of different voltage level batteries and Table 4.10 compares the different types of batteries.

The types of batteries used in distribution substations are as follows.

1. Flooded lead–acid
2. Flooded nickel–cadmium
3. AGM type lead–acid VRLA
4. Gel type lead–acid VRLA

**Table 4.9** Different voltage levels of batteries

| S. No. | Nominal voltage rating of batteries (V) | Nominal single cell voltage (V) | Float cell voltage (V) | Number of cells | Permissible DC voltage variation (V) | End of discharge cell voltage (min) (V) |
|---|---|---|---|---|---|---|
| 1. | 24 | 1.2 | 1.4–1.42 | 19 | 21.7–27 | 1.14 |
| 2. | 30 | 1.2 | 1.4–1.42 | 23 | 26.2–32.7 | 1.14 |
| 3. | 110 | 1.2 | 1.4–1.42 | 87 | 99.2–123.5 | 1.14 |
| 4. | 220 | 1.2 | 1.4–1.42 | 170 | 193.8–241.4 | 1.14 |

**Table 4.10** Comparison between different types of batteries

| S. No. | Flooded Lead–Acid | Flooded Nickel–Cadmium | VRLA batteries |
|---|---|---|---|
| 1. | Battery technology is more reliable for motionless application (i.e., substation in transmission and distribution network) | An excellent battery for floating, cycling and engine | Low initial investment cost |
| 2. | Expensive | Starting applications especially in extreme temperature environments | No need for spill containment |
| 3. | Need maintenance: no good without proactive maintenance battery performance | Environmental concern regarding cadmium toxicity | Smaller floor requirements |
| 4. | Need ventilated and temperature-controlled room | High initial investment, poor lifecycle leads to high cost | Higher power density |
| 5. | Need spill containment, occupy large footprint | Ni–Cd battery works in harsh temperature environment and has lesser lifecycle cost compared to lead–acid batteries | No need for separate room and safer for maintenance |
| 6. | Not the best choice for distribution substations | – | Poor lifecycle due to technology limitation such as corrosion, negative polarization, thermal run-away, and dry out |
| 7. | – | – | Require temperature-controlled environment |
| 8. | – | – | Require extensive maintenance and monitoring |
| 9. | – | – | Very sensitive to improper charging |

5. Nickel–cadmium VRLA
6. NB: VRLA (valve regulated lead–acid) batteries.

### *4.10.2 Common Causes of Fault and Best Practices for Battery Maintenance*

1. **Matching the Charger to Battery Requirements**: Poor battery charging practice makes its life shorter, and as the cost of the battery is expensive it should be preserved carefully. A good choice of charging practice will lead to

longer and healthier battery life. The conditions to be taken care of in the battery are how many hours it should be charged, how it should be discharged, temperature limits in battery operation, and the required number of cells maintained depending on the application.

2. **Constant Voltage Charging**: In the battery charging process the constant-voltage region mode is the event where the charging battery is in constant input voltage irrespective of battery state of charge. There is more initial current to the battery during this event because of the superior potential difference between the battery and charger. This approach helps the battery to charge at a faster pace.
3. **Avoiding Overdischarge**: When the battery is fully discharged it is removed from the serving load; it would increase the lead–acid batteries. The typical cutoff voltage of a lead–acid battery is 1.75 V; this voltage is a critical parameter related to battery operating temperature and discharge rate. However, overdischarging may affect battery life as well as the recharging process. Also, overdischarging may cause lead to be precipitated in the separator and cause a short in the cell or between cells.
4. **Cleaning**: Cleanliness means being clean and dirt-free, which is applicable for maintaining the long lifetime of the battery. This would reduce expenses of battery repair due to corrosion and dirt. Generally dry dirt is deposited in the battery during its normal operation; it may cause blow-off or explosion because this dirt makes stay current a conductor.
5. **Avoiding High Temperature**: Temperature is a harmful parameter of the battery, which is to be maintained in a low range of 55 °C. The corrosion rates, metal component, chemical catalyst process, and self-discharges in the battery will increase its temperature.
6. **Safety Precautions**: Safe precautions should be maintained in lead–acid batteries because they may expel hazardous sulfuric acid; release of hydrogen and oxygen would cause a severe explosion and may also cause generation of toxic gases such as arsine and stibine. These problems are conquered by proper adoption of precautions such as wearing a face mask, aprons, and gloves when working with battery acid components. If battery acid accidentally gets into the eyes, it is highly recommended to flush the affected area with clean water followed by proper medical intervention.

**Table 4.11** Preventative maintenance schedule of battery and battery charger

| Inspection frequency | Items to be inspected | Inspection notes | Action required for unsatisfactory conditions |
|---|---|---|---|
| Weekly | Battery voltage | Battery voltage will be recorded for each battery set from the corresponding charger with switch ON battery position | – |
| | Electrolyte level | Check the float level indicator | If electrolyte is below the level, top up the cell with pure distilled water by using plastic funnel |
| | Exhaust fan | Exhaust fans should be adequate and in running condition | If any fan is out of order, it is to be repaired/replaced if necessary |
| | Voltage of pilot cells | Voltage will be recorded with the help of voltmeter keeping the flat charger in ON position | Pilot cells should be selected in such a manner that all cells of the set are checked within a week |
| | General cleanliness | Battery sets, stands, and general battery room is properly clean and dry. No extra material should be kept in battery room | – |
| Monthly | Cell voltage | Voltage of all cells recorded with voltmeter keeping the charger in ON position and in FLOAT condition, not on BOOST charge | Replacement of contact |
| | Specific gravity | Specific gravity measured with hydrometer | – |
| | Ventilation plug | Check that filling plugs are not blocked in any cell | – |
| | Connection | Check connections for tightness and corrosion | Old jelly to be replaced after thorough cleaning |
| | Leakage | Visually check cells for electrolyte leakage and crack | If cracks or leakage, replace cell |
| | Indication and fuses | Check battery chargers for healthy fuses and indications. (1) Carry out equalizing charging of battery. (2) Check operation of DC emergency light. (3) Check there is no switch in the battery room that can cause a spark and so on | – |
| Half-yearly | Condition of cable trenches | Check sand in cable trenches and rubber packing where cables are passing through trench cover | Provide sand and rubber packing if required |

(continued)

**Table 4.11** (continued)

| Inspection frequency | Items to be inspected | Inspection notes | Action required for unsatisfactory conditions |
|---|---|---|---|
| | Light and heaters | Check the light and heaters inside the charger | Replace if damaged or defective |
| Yearly | Charger cleanliness | Earthing connection should be checked at least once a year by Meggering the resistance level. The same should be recorded and preventive steps should be taken | – |
| | Condition | Check all connections in the battery charger for tightness of insulators and so on | Tighten loose connections and provide insulation where required by insulating tape |
| | Charger cleanliness | Clean charger from inside with vacuum cleaner to ensure complete removal of dust, cobwebs, and so on | – |
| | Condition of insulation | Visually check physical condition of cell stand insulators | Replace damaged insulators |
| | Condition of battery stands leads, discs, rubber pad, and so on | Check battery stands for physical condition of rubber pads, leads, and discs required for leveling of cells. Also check condition of paint | Repair stands if necessary. Replace damaged rubber pads/lead discs and also repaint the stands with acid -proof black paint if required |
| | Conditions of protective clothing | Check physical condition of the apron, rubber gloves, and so on | If damaged replace the same |
| | Condition of acid -proof painting | Visually check condition of acid-proof paint inside battery room | Repaint if required |
| | Standardization of hydrometer | Check hydrometer for standardization by comparing its reading with threading of same cell with a new/standard hydrometer | If there is difference, replace the hydrometer |
| | Exhaust room | Check condition of exhaust fans | Overhaul the exhaust fans and replace those defective/damaged |
| | Lighting in battery room | Check the light fittings of battery room | Replace if damaged or defective |

### 4.10.3 *Preventive Maintenance of Battery*

A preventive maintenance schedule is shown in Table 4.11.

## 4.11 Earth Grid and Earthing System

One of the important aspects in the operation of protective equipment is proper earthing. Earthing is the process of increasing reliability of the power supply; it also helps to provide a stable voltage profile and prevent voltage fluctuations during a fault event. For an outdoor substation, a main earthing ring should be provided around the substation which should be connected to all earth electrodes [19]. The ring should be laid so as to have the shortest connection from transformers, circuit breakers, and the like.

### 4.11.1 *Types of Earthing*

It can be divided into neutral and equipment earthing.

1. Neutral earthing deals with the earthing of the system being neutral to ensure that neutral points are held at earth potential and a return path is available to the neutral current. Points to be earthed are: the transformer neutral is to be earthed to two separate and distinct earth electrodes interconnected with the substation earth mat.
2. Equipment earthing deals with earthing of noncurrent-carrying parts of equipment to ensure safety to personnel and protection against lightning. Points to be earthed are: All noncurrent-carrying metallic parts of equipment, structures, enclosures, overhead shielding wires, bushing flanges, transformer cores, cable sheaths, earthed screens, pipes, portable appliances, fences, doors, and screens.

### 4.11.2 *Common Earth System for Low- and High-Voltage Systems*

In a substation a common earth bus is used for both high as well as low-voltage systems. If the low- voltage neutral is not connected to the common earth system but has a separate earth bus, there will be a difference of potential between the high- and low-voltage neutrals and a dangerous potential gradient across the earth surface can exist that can endanger life [20]. With a low-resistance earth bus and the neutrals connected to a common earth system, there will be no danger to the

**Table 4.12** Preventive earth maintenance schedule

| Inspection frequency | Items to be inspected | Inspection notes | Action required for unsatisfactory conditions |
|---|---|---|---|
| Weekly | Cleaning of earth pit | Earth pit needs to be cleaned and free from garbage | – |
| | Water in earth pit | Salt or plain water to be poured in earth pit at least once a week | – |
| Monthly | Check connectivity between electrodes | Connectivity between earth electrode and earth grid to be checked | In case of failure to get the connection it has to be replaced with new part |
| | Physical health check of earth electrode | Physical condition of earth electrode has to be checked and corrected if required | Electrode needs to be replaced if found broken |
| Half-yearly | Measurement of earth resistance | It is important to measure the earth resistance at regular intervals. If found to be poor it should be corrected by replacing the same or adding new pits | – |

low-voltage system and keeping everything in the station at a common potential above earth will outweigh the disadvantages. Table 4.12 gives a maintenance schedule for the system.

1. **LA Earthing**: The earthing lead for any LA will not pass through any iron or steel pipe, but will be taken as directly as possible from the LA to a separate earth electrode interconnected with the substation earth mat. Individual earth electrodes should be provided for each station-type lightning arrester, whereas for a distribution type lightning arrester, one electrode may be provided for a set of lightning arresters.
2. **Coupling Capacitor Earthing**: A separate earth electrode, generally a driven rod or pipe, should be provided immediately adjacent to the structure supporting the coupling capacitors of carrier current equipment. This earth should only be used for high-frequency equipment.

    1. **Overhead Line Earthing**: (i) Line conductors and insulators by ensuring the operation of the protective control gear under such conditions; (ii) to discharge lightning strikes to earth. (iii) to minimize inductive interference with the communication circuits. One or more earth GI wires are run along the power line (above the conductors).

## References


1. Pabla AS Electric power distribution, 6th edn. Tata McGraw Hill, New Delhi
2. IEEE Surge Protection Standards Collection (1992) Edition
3. Fink DG, Beaty HW Standard handbook for electrical engineering, 12th edn. McGraw Hill, New York
4. Hornak D, Zipes D (1991) Automated bus transfer control for critical industrial processes. IEEE Trans Ind Appl (Sept/Oct)
5. Wagner VE, Andreshak AA, Staniak JP (1990) Power quality and factory automation. IEEE Trans Ind Appl, pp 620–626 (July/Aug)
6. Manual on Transmission Criteria, Central Electricity Authority (1985) p 15 (May)
7. IS: 1255-1983, Code of practice for installation and maintenance of power cables up to and including 33 kV
8. Miller TJE (1982) Reactive power control in electrical system. Willey, New York, pp 353–361
9. Pabla AS (1985) Distribution system reactive power. IEMA J, pp 11–18 (July)
10. Lee WH (1997) Power distribution planning reference book. Marcel Dekker, New York
11. Jayaraman B (1993) Earth resistance improvements. J Inst Eng (India), pp 39–40 (April)
12. Power Sub-transmission and Distribution Systems-Prevailing practices in India (1987) CBI&P, Publication No. 198, New Delhi, December, pp 75–79
13. IS: 5613 (Part I and II), Design installation and maintenance of lines up to and including 220 kV
14. Tagare DM (2002) Electrical power capacitors—design and manufacture. Tata McGraw Hill, New Delhi, pp 207–301
15. IEEE Protective Relaying System Standards Collection (1991) Edition
16. Bozic Z, Hobson E (1997) Urban underground network expansion planning. In: IEEE proceeding on generation, transmission and distribution, vol 14, no 2, pp 118–124 (March)
17. Short TA (2004) Electric power handbook. CRC Press, Boca Raton, p 689
18. IEEE Battery Standards Collection (1991) Edition
19. Earthing System Parameters for HV, EHV and UHV sub-stations (1985) Technical Report 49, CBI&P, New Delhi (Sept)
20. Grounding Systems and Practices (1989) Central Power Research Institute, Technical Report No. 182

# Chapter 5
# Best Practices in Operation and Maintenance of Transformers

## 5.1 Power Transformer

### *5.1.1 Description*

The transformer (Fig. 5.1) is static and is one of the most important devices in a power system network. It does step-up or step-down of the voltage and transfers power from one circuit to other AC circuits without any change in frequency. The transformer has primary and secondary windings housed in the main tank [1]. The transformer capacity is expressed in volt–ampere (kVA or MVA).

The transformers comply with IS0: 2026; fittings and accessories comply with IS 3639. A transformer of 33/11 kV rating should have a delta star winding connection. The favored ratings for 33/11 kV transformers should be 6.3, 8, 10, 16, 20 MVA for urban areas and 1, 1.6, 3.15, 5, 6.3 MVA for rural areas. Any standard ratings other than stated above may also be chosen depending upon technical and economic considerations.

The transformer ventilation can be of dry type, epoxy cast dry-type, or oil-filled depending on whether transformer installation is indoor or outdoor. The cooling should be ONAN or ONAF. The core is usually constructed with high-grade rolled grain-oriented steel. Transformers should withstand combined voltage, frequency, and fluctuations without injurious heating and produce over-fluxing conditions such as 125% for 1 min and 140% for 5 s. The tolerable increase in temperature rise of winding and oil should be 35 and 40 °C, respectively, over an ambient temperature of 50 °C.

The transformers should be provided with relays such as oil and gas actuated relays connected with trip contacts and alarm, if applicable. The tap changing switch is placed at a well-suited position for easy operation of the switch from ground level. For locking the switch, the handle equipped with locking is set up along with the tap position indication. For better voltage control of transformers of



J. Raja et al., *Practices in Power System Management in India*,
Power Systems, https://doi.org/10.1007/978-981-10-2972-1_5

**Fig. 5.1** Power transformer

3.15 MVA and above, the tap changing device in the onload circuit should be provided. Transformers with an onload tap changing device should have taps in steps of 2.5% ranging from (+)5% to (−)15% voltage variation on the 33 kV transformer winding [2].

### 5.1.2 *Causes and Nature of Failure*

In normal operation the power transformer is protected from various kinds of relays and protective devices. However, during analysis the reasons for failure of a power transformer mainly concentrate on the following.

- Oil degradation/leakage in the transformer tank
- Neutral shifting/disturbance due to failure of earthing or imbalance of the phasewise current
- Internal short-circuit due to failure of insulation inside transformer
- Increase in winding temperature due to internal short-circuit
- Failure of protective devices including temperature indicator, relay settings, and the like
- Overloading of power transformer without proper fuses/protective devices to cut down the load
- Fluctuation of voltage at incomer side
- Loose contact between live parts of the distribution system.

### 5.1.3 Primitive Maintenance of Power Transformer

Primitive maintenance to be taken depends upon the size of the transformer, operation of the protection relay, whether tripping of the relay makes a loud noise, smoke, or discharge of oil from the transformer, and so on. Visually inspect a transformer's physical condition for any type of damage to the bushings. Check the temperature of oil and oil level in the conservator tank. Use Megger for taking readings between the primary and secondary of the transformer and also of the individual side of windings to earth. If everything goes right, proceed as follows.

The failure may possibly be caused by sudden and heavy overload, or short-circuit. Check if its ampere rating is right or the fuse has dropped out; if wrongly placed, replace by the right size and energize the transformer, after switching off the load. If everything seems right, close the secondary circuit; if the fuse blows up again, the fault is surely in the outgoing lines, which should be mapped and rectified if possible. On the other hand, if the primary circuit fuse blows up, even when the load is not connected, an internal fault is indicated. This condition also applies, but only if an overcurrent relay has operated and tripped the circuit breaker.

When a transformer is first switched on and if a differential relay operates, it may be due to a switching surge. Check the harmonic control circuit and settings. However, if the relay operates when the transformer is in operation, it is surely because of an internal fault.

Any kind of tripping of the Buchholz relay requires careful examination. If the upper assembly of the relay has operated due to slow evolution of gas it is necessary to find its composition before jumping to conclusions. If, on the other hand, the lower relay assembly has tripped due to sudden release of a large quantity of gas, a major internal fault is to be surmised especially if differential or overcurrent or earth fault relay has operated. If it is only air, there is no reason for worry, as air can enter the transformer in many ways.

Thorough checking is necessary if there is an evolution of smoke or oil, or the earth fault relay has tripped or not, and also PRV has operated in the case of a large transformer. In such cases, reclosing the circuit breaker should not be permitted as it may cause further damage. Detailed testing of the transformer is to be carried out and the results compared with test certificate figures and the manufacturer consulted. In most cases, the cause of the fault can be found if the condition of the windings is carefully observed by lifting the core and coil assembly. The following explanations will help to detect the cause.

**Lightning discharge or overvoltage**: The occurrence of breakdown happens at the coil end turn near the line terminal. Also it has the possibility of appearing as a breakdown or flashing in the turn of the oil or lead end terminal and close-earthed part whereas the rest of the coil is healthy.

**Sustained overloads**: The sign of excessive heating present in one of the windings or all the phases would lead to weak insulation and it may lose its physical property.

**Interturn short**: As in sustained overloads similar signs can be observed, but this would appear on only one coil and rest is integral.

**Dead short-circuit**: The wrong coil displacement is distinguished in the dead short-circuit. The coil is twisted in loose fashion and also may cause an explosion under tension conditions.

### 5.1.4 Visual Checking of Transformer

- Check the silica gel color. If it is pink, reactivate or replace it. Also ensure proper quantity of oil in breather oil cup.
- Check oil level in conservator of main tank and OLTC. It should be >½ level marking.
- Check oil level in bushings.
- Check for any oil leakage and arrest if any.
- Check the working of Oil Temperature Indicator (OTI) and Winding Temperature Indicator (WTI) by taking hourly temperature readings. There should be changes in readings as per loading of transformer and atmospheric conditions.
- Check the cooling system by manually operating fans/pumps.
- Check the tap position of remote tap changer control (RTCC) and on-load tap changer (OLTC) panels. It should have the same position number.
- Check the humming noise and vibration of the transformer. If any abnormality is found, it is to be referred to the manufacturer concerned.

### 5.1.5 Preventive Maintenance Schedule of Power Transformer

The following three factors directly affect the life of the distribution transformer.

**Excess heat**: Excess heat, on account of overloading, loose connections, low oil level, blocking in oil circulation, and so on.

**Moisture**: The oil dielectric strength is reduced once moisture enters the transformer through the breather and contaminates the transformer oil.

**Oxygen in the air**: It oxidizes the oil to form sludge and acids that attack the insulation and make it brittle; it corrodes the surface of the inner and conservator tanks.

Maintenance work on the distribution transformer is intended to take care of the above three factors. The periodic maintenance checks to be carried out are given in Table 5.1.

**Table 5.1** Condition monitoring of transformer

| S. No. | Components | Test results | Condition evaluation | Remarks and required action |
|---|---|---|---|---|
| 1. | Paper insulation on windings | • Recovery voltage meter (moisture content)<br>• Tan-delta value<br>• DGA (violation of standard value of different gases)<br>• Insulation resistance value<br>• Furan content | Whether paper insulation is dry or wet | The following test will detect whether drying out is necessary Internal checkup is required or not |
| 2. | Transformer oil | • DGA test<br>• Moisture content in oil<br>• Color | Quality of oil is good or not | If it is bad, then replacement is required |
| 3. | Winding mechanical integrity | FRA test vibration to be assessed | Mechanical integrity of winding is satisfactory. Clamping pressure and core tightness of winding is normal or not | Any repair is possible or not |
| 4. | Transformer core | • DGA test<br>• FRA test<br>• Vibration assessment ratio test | Transformer core, core insulation, and core tightness are good or not | Any repair or replacement is required or not |
| 5. | Bushings | • HV bushing tan-delta value<br>• LV bushing tan-delta value | Whether the bushings are healthy or not | If not healthy, capacitance and tan delta test is to be performed again using the automatic test kit |
| 6. | OLTC | • Winding resistance at each tap<br>• Ratio test<br>• Oil color check<br>• Operation check | Whether it is healthy or not | If found unhealthy. this is to be repaired for rectification |
| 7. | Other components | • PRD fine and fit<br>• OTI or WTI fine calibration<br>• Radiator banks: no blocking, multiple point leakages<br>• Buchholz: healthy<br>• Values: healthy, no leakages | All are healthy or not | If radiator bank is not in good shape, it is better to replace the bank. If PRD is not found healthy, modified kit is to be put on PRD to avert possible false tripping due to ingress of moisture |

(continued)

**Table 5.1** (continued)

| S. No. | Components | Test results | Condition evaluation | Remarks and required action |
|---|---|---|---|---|
| | | • Main gasket of transformer: in good shape<br>• Gasket of MBS: in bad shape<br>• Turrets: healthy<br>• Cabling: some cables are cracked | | Damaged gaskets and cracked cables are usually replaced with the new one |
| 8. | General appearance | • Painting: check for repainting<br>• Oil leakages: check for multiple leakage<br>• Terminal connection: analyzed by thermovision scanning All foundations are checked properly. Analysis is also done to check if there are any requirements for oil pit cleaning | Whether the general appearance is good | The unsatisfactory cases are to be reanalyzed |

### 5.1.6 *Predictive Maintenance*

**Buchholz Relay**: This relay is designed to detect a transformer internal fault in the initial stage to avoid major breakdown. An internal fault in the transformer generates gases by decomposition of oil due to heat and spark inside the tank. These gases pass towards the conservator tank, trapped in relay, and thereby causing the oil level to fall. The upper float rotates and the switches' contacts close, thereby giving an alarm signal. In the case of a serious fault, there is more gas generation, which causes operation of the lower float and trips the circuit breaker. The gas can be collected from a small valve fitted at the top of the relay for dissolved gas analysis (DGA).

**Dissolved Gas Analysis (DGA)**: The transformer, in operation, is subjected to various thermal and electrical stresses, resulting in liberation of gases from the oil used as insulation media and coolant. The solid insulating materials such as paper, wooden support, and pressboard cause degradation and form different gases, which are dissolved in the oil. The most significant gases generated are hydrogen ($H_2$), methane ($CH_4$), ethylene ($C_2H_4$), acetylene ($C_2H_2$), propane ($C_3H_8$), propylene ($C_3H_6$), carbon monoxide (CO), carbon dioxide ($CO_2$), and ethane ($C_2H_6$). The gas connected in the relay will help to identify the nature of the fault. The greater the rate of gas collection, the more severe is the nature of the developing fault.

### 5.1.7 Condition Monitoring of Transformer

Based on the assessment of various test results, the health of the various components is encapsulated as shown in Tables 5.1 and 5.2.

## 5.2 Distribution Transformer

### 5.2.1 Description

The transformer should be in conformance with IS-1180, IS 2026, and IS-11171. Distribution transformers can be 33/0.4 kV, 22/0.4 kV, or 11/0.4 kV. The 33/0.4 kV distribution transformers generally have standard ratings of 500, 630, 1000, 1250, 1600, or 2000 kVA depending on necessities. In rural and urban areas where the population is not large lower rating transformers can be employed [3]. The usual standard ratings of 11/0.4 kV distribution transformers are given as 6.3, 10, 16, 25, 50, 63, 100, 250, 315, 400, 500, 630, and 1000 kVA.

The efficiency of 11/0.4 kV distribution transformers of 100 kVA and below should not be less than 98% at both 50 and 100% loading. The efficiency of 11/0.4 kV distribution transformers above 100 kVA should not be less than 98.8% at both 50 and 100% loading. The efficiency of 33/0.4 kV distribution transformers should not be less than 99% at both 50 and 100% loading.

The transformer can be classified as oil-filled or dry-type based on the requirements. Materials such as high-grade cold rolled grain-oriented (CRGO) steel and amorphous material are employed in the construction of energy-efficient and energy-saving types of instruments. The scrap CRGO materials are not incorporated in the construction of the transformer. Low-capacity transformers of rating 100 kVA and less should normally be used. The higher capacity transformers of rating larger than 100 kVA should be used where there are space constraints such as concentrated loads or areas with a high density of load.

**Table 5.2** Gas color for finding affected material

| S. No. | Color | Identification |
|---|---|---|
| 1. | White | Decomposed paper and cloth insulation |
| 2. | Yellow | Decomposed wood insulation |
| 3. | Grey | Overheated oil due to burning of iron portion |
| 4. | Black | Decomposed oil due to electric arc |

**Table 5.3** Interpretation of the faults according to the observed ratios of gases

| S. No. | Characteristic Fault | Ration code | | | Diagnosis |
|---|---|---|---|---|---|
| | | $C_2H_2/C_2H_4$ | $CH_4/H_2$ | $C_2H_4/C_2H_6$ | |
| 1. | No fault | 0 | 0 | 0 | Regular ageing |
| 2. | Partial discharge of low-energy density gas | 0 | 1 | 0 | Discharge in gas-filled cavities due to incomplete impregnation |
| 3. | Partial discharge of high-energy density gas | 1 | 1 | 0 | As above but leading to tracking or perforation of solid insulation |
| 4. | Discharge of low-energy gas | 1–2 | 0 | 1–2 | • Continued sparking in oil between bad connections of different potential or to floating potential<br>• Breakdown of solid material |
| 5. | Discharge of high-energy gas | 1 | 0 | 2 | Discharge of power followthrough arcing. Breakdown of oil between winding or between coil to earth |
| 6. | Thermal fault of low temp less than 150 °C | 0 | 0 | 1 | General insulated conductor overheating |
| 7. | Thermal fault temp 150–300 °C | 0 | 2 | 0 | Flux influence causes overheating of core, increases hot spot; pyrolitic carbon deposition makes joints bad contacts and circulation of current over tanks |
| 8. | Thermal fault temp 300–700 °C | 0 | 2 | 1 | |
| 9. | Thermal fault >700 °C | 0 | 2 | 2 | |

Depending on the type of supply transformer there can be three phases or a single phase. The ONAN cooling method is used for oil-filled transformers. Based upon technical and economic considerations, any standard rating other than the above could also be chosen.

Faults are shown in Table 5.3.

## 5.2.2 *Transformer Maintenance Management*

The major types of fault causing transformer failure are as follows.

- Insulation failure
- Design/material/workmanship
- Oil contamination
- Overloading
- Line surge
- Improper maintenance/operation

- Loose connection
- Lightning/flood
- Moisture.

### 5.2.3 *Causes of Failure of Distribution Transformers*

Table 5.4 lists some of the causes of failures of transformers in service and probable corrective measures have been highlighted to reduce the rate of failure, which are primitive alarming signals in current modern-day distribution transformer networks across the country [4, 5].

**Table 5.4** Causes of DT failure

| S. No. | Reason on account of user | Probable cause | Preventive action |
|---|---|---|---|
| 1. | Prolonged overloading | Selection of higher rating fuses | • Frequent measurement of peak load currents through tongue tester and fixing the correct rating of fuses on both HT and LT side. If overloading is 20% during peak load and persistence, then propose new transformer/ augmentation<br>• Use correct size of fuse wire. Utilities should make the fuse wires available. Avoid using two wires in the fuse |
| 2. | Unbalanced loading | Single-phase loading or additional heat on account of unbalanced current | • Avoid single-phase loading and unbalanced loading of distribution center. Measure and record the neutral current at regular intervals in distribution center history card<br>• If neutral current exceeds 10% of load current, remedial measures should be taken to bring the neutral current down to within acceptable limits |

(continued)

**Table 5.4** (continued)

| S. No. | Reason on account of user | Probable cause | Preventive action |
|---|---|---|---|
| 3. | Faulty terminations | Once spark strikes the cable terminal because of loose connections it causes damage to busing sealing gaskets resulting in oil leakage on bushing top and causing failure of distribution transformer due to low oil level | Use proper termination connectors at both HV and LV sides by bimetal strips per ISS and REC standards |
| 4. | Power theft and hooking | Results in overload/ unbalanced load and may cause failure in due course | • Meter the DTC to identify the pilferage<br>• Regular energy audits, surprise raids, use of ABC conductor |
| 5. | Wrong earth connections | High resistance will delay fault clearance | Solid earth is extremely essential. Earth at two diagonally opposite points of tank. Ensure that nut bolts are not painted at earthing |
| 6. | Prolong short-circuit | If the fault persists for longer duration will result in melting, brazing, or brining of winding | Necessary protection devices may be installed properly such as OCB on HV and ACB on LV side; appropriate rating of HRC fuses for small transformers |
| 7. | Less or no maintenance | – | Line workers should prepare their own maintenance charts as per the criticality (i.e., transformer rating, connected loads, type of installation, and place) |
| 8. | Poor quality of LT cables and hanging against the terminal bushings without any support | PVC insulation causing dead short-circuit in the transformer. Also, its weight pulls the terminal down which may affect the leakage oil from the sealing gaskets of the bushing | – |
| 9. | Improper installation of distribution center | No adequate clearance, missing breather, tree branches near terminals, explosion vent diaphragm, dielectric oil value of oil, and so on | Necessary precaution during replacement/ installation of transformer |

### 5.2.4 Preventive Maintenance Schedule

The schedule is shown in Table 5.5.

**Table 5.5** Transfer maintenance schedule

| S. No. | Inspection frequency | Items to be inspected | Inspection checkpoint | Action required for unsatisfactory conditions |
|---|---|---|---|---|
| 1. | Monthly | • Ambient temperature | • Note ambient temperature regularly | – |
| | | • Oil level in transformer | • Check against transformer oil level | • If transformer oil level is low refill the oil and examine transformer |
| | | • Relief diaphragm | • Check physical condition | • Replace if cracked or broken |
| | | • Dehydrating breather | • Check that air passages are free. Check color of active agent | • If silica gel is pink, change by share change. The old gel may be reactivated for reuse |
| 2. | Quarterly | • Bushing | • Examine for cracks and dirt deposits | • Clean and replace |
| | | • Load (amp) and voltage | • Check against rated figures | • Shut down the transformer and investigate if current and voltage are higher than the normal value |
| | | • Oil and winding temperature | • Check temperature rise is reasonable | • Shut down the transformer and investigate if the temperature is higher than the normal value |
| | | • Oil in transformer | • Check for dielectric strength and water content | • Take suitable action |
| | | • Cooler fan bearings, motors, and operating mechanism | • Lubricate bearing. Check gear boxes. Examine contacts. Check manual control and interlocks | • Replace burnt or worn contact or other parts |
| | | • OLTC | • Check oil in OLTC driving mechanism | – |

(continued)

**Table 5.5** (continued)

| S. No. | Inspection frequency | Items to be inspected | Inspection checkpoint | Action required for unsatisfactory conditions |
|---|---|---|---|---|
| 3. | Half-yearly | • Earth resistance | • Measure earth resistance and record | • Replace with new earthing |
| | | • IR value of lightning arrester | • Measure insulation resistance of lightning arrester | • Change lightning arresters immediately |
| 4. | Yearly | • Oil in transformer | • Check for acidity and sludge | • Filter or replace |
| | | • Oil-filled bushing | • Test oil | • Filter or replace |
| | | • Gasket joints | – | • Tighten bolts evenly to avoid uneven pressure |
| | | • Cable boxes | • Check for sealing arrangements for filling holes | • Replace gasket, if leaking |
| | | • Surge diverter and gaps | • Examine for cracks and dirt deposits | • Clean and replace |
| | | • Relay, alarms, and control circuits | • Examine relays and alarm contacts, their operation, fuse, and so on. Test relays | • Clean the component and replace contracts and fuses if required |
| | | • AB switch | • Check the operation of AB switch | • Grease and check the operation |
| 5. | Two yearly | Sampling and testing of acidity of oil | Take the sample and test for dielectric strength | Replace transformer oil |
| 6. | Three yearly | Filtering of oil (Class B or mixture of Class A and B) | Take the sample and test for dielectric strength | Replace transformer oil |
| 7. | Five yearly | Overhaul of transformer | | Replacement every fifth year |

## 5.3 Operation and Maintenance of Transformers and Reactors

### *5.3.1 Power Transformers and Shunt Reactors*

Power transformers and reactors are critical components of the modern electrical power system. To provide reliable, extensive, and continuous services, it is very crucial that regular operation and maintenance of transformers/reactors and their parts are well planned and carried out. The frequency of O&M is dependent on environment, climate, service condition, and so on. A rigid preventive maintenance ensures the long life of transformers/reactors, as well as trouble-free service and low maintenance cost [6]. Table 5.6 lists the maintenance schedule of transformers/ reactors and maintenance consists of regular inspection, testing, and reconditioning where necessary.

**Table 5.6** Recommended maintenance schedule for transformers

| S. No. | Inspection frequency | Items to be inspected | Action required in case of undesired condition |
|---|---|---|---|
| 1 | Hourly | • Ambient temperature<br>• Winding temperature<br>• Oil temperature<br>• Load (amperes)<br>• Voltage<br>• Tap position of tap changed | • To be compared with designed value<br>• Refer to OEM in case of violation persistently higher than normal |
| 2 | Monthly | • Oil level in bushing, conservator, and OLTC conservator<br>• Manual starting of oil pumps and fans<br>• Checking of oil leak<br>• Oil level in breather oil seal cup<br>• Condition of silica gel<br>• Changeover of supply in MB (to be checked manually if two parallel AC supplies are provided for auxiliary) | • Bushing needs to be replaced in case of oil leak to avoid any failure. In case of low oil level in main conservator or OLTC conservator, top up with dry oil at the earliest opportunity<br>• In the case of oil leak, check tightness of bolts, cracks, and damage to oil sealing gaskets. In the case of a damaged gasket, gasket needs to be replaced<br>• Regenerate or replace the silica gel if it is saturated/moist |
| 3 | Half-Yearly | • Dissolved gas analysis of oil<br>• Oil parameter for BDV, PPM, tan-delta, resistivity, IFT<br>• Operational checks on OLTC in charged but offload condition for all tap positions | • In case of violation of fault gases or oil parameter beyond the specified limit, monitoring schedule can be changed in consultation with OEM or based on the criticality of the transformer/reactor |
| 4 | Yearly | • Auto starting of fan and pumps<br>• Measurement of BDV of OLTC oil<br>• External cleaning of radiators, bushings, Buchholz, and gas collecting device, and so on<br>• Maintenance of OLTC, driving mechanism, transformer MB<br>• Online moisture measurement<br>• Rust, damage, and repairing (if required)<br>• Core insulation test<br>• Bushing capacitance and tan-delta | • If any problem is encountered in OLTC such as OLTC balancing, problem in driving mechanism, same needs to be rectified per the procedure or as recommended by OEM |

(continued)

**Table 5.6** (continued)

| S. No. | Inspection frequency | Items to be inspected | Action required in case of undesired condition |
|---|---|---|---|
| 5 | Two yearly | • OTI and WTI setting checks<br>• Alarm and trip test of WTI,OTI, PRD, Buchholz, MOG, and SPR<br>• Earth pit resistance | – |
| 6 | Four yearly | • Winding capacitance and tan-delta<br>• Calibration of OTI and WTI | |
| 7 | Ten yearly | • OLTC overhauling | |
| 8 | SOS | **All LV tests including**<br>• Winding resistance measurement of winding<br>• Magnetic balance test<br>• Magnetizing current test<br>• Voltage ratio at all taps<br>• Frequency response analysis<br>• Measurement of PI and DAI<br>• Short-circuit impedance test<br>• NGR winding and bushing capacitance and tan-delta<br>• Vibration measurement for reactors<br>• Furan measurement | Test results to be compared with test/results of transformer before its commissioning. Previous results to be used in case of major deviation |
| 9 | 25 Years | • Residual life assessment of transformer/reactor | – |

### 5.3.2 *Testing of Transformers/Reactors*

Testing is shown in Table 5.7.

### 5.3.3 *General Maintenance of Transformer/Reactor Accessories*

**Silica Gel Breather Check**: Breathers should be examined to ascertain if the silica gel requires changing. For better results it is advised to change the silica gel when half to two-thirds of the silica gel has become saturated [7]. The silica gel can be collapsed into a shallow tray or reactivated at the same time as in its charge container. Heating of the gel should be done in a well-ventilated oven at 110–130 °C for 8 to 10 h or 150–200 °C for 2 to 3 h and can be used again. Saturation of silica gel charge must occur from the bottom. In the case of silica gel desiccant saturated from

**Table 5.7** Transformer/reactor health

| S. No. | Name of test/checkpoint | Purpose of test/check |
|---|---|---|
| 1. | Core insulation tests | Allows for investigating accidental grounds that result in circulating currents if there is more than one connection between the core and ground |
| 2. | Earth pit resistance measurement | To check the resistance of earth pit provided for transformer. Proper treatment is to be given if the resistance of earth is more |
| 3. | Insulation resistance (IR) measurement | Test result shows insulation strength, degree of paper insulation degradation, chemical adulteration in the oil, and damage inside the transformer tank |
| 4. | Capacitance and tan-delta measurement of bushings | Measurement of C1 and C2 capacitance and tan-delta in UST mode. Any change in value of normal insulator capacitance indicates nonstandard conditions such as open circuits or short-circuits in the capacitance network, presence of moisture layer |
| 5. | Capacitance and tan-delta measurement of windings | With the measurement of loss/dissipation factor and capacitance of winding the usual condition of the ground and interwinding insulation can be detected |
| 6. | Turns ratio (voltage ratio) measurement | Measurement used to recognize any abnormal conditions in tapping changer or in shorted or open turns and so on |
| 7. | Vector group and polarity | To determine the phase relationship and polarity of transformers |
| 8. | Magnetic balance test | This test confirms any sort of variance in the magnetic circuit and is applicable only on three-phase transformers |
| 9. | Floating neutral point measurement | This test is conducted to ascertain possibility of short-circuit in a winding |
| 10. | Measurement of short-circuit impedance | This test is used to identify winding movement that usually occurs due to heavy fault current or mechanical damage during transportation or installation since dispatch from the factory |
| 11. | Exciting/magnetizing current measurement | Measurement used for shifting of windings, locating any fault in magnetic core structure, problems in tap changers, or failures in turn-to-turn insulation |
| 12. | Operational checks on OLTCs | To ensure smooth and trouble-free operation of OLTC |
| 13. | Tests/checks on bushing current transformers (BCTs) | To ascertain the health of bushing current transformer at the time of erection |
| 14. | Operational checks on protection system | Operational checks on cooler bank (pumps and fans), breathers (silica-gel or drycol), MOG, temperature gauges (WTI/OTI), gas actuated relays (Buchholz, PRD, SPR, etc.), and simulation test of protection system |

(continued)

**Table 5.7** (continued)

| S. No. | Name of test/checkpoint | Purpose of test/check |
|---|---|---|
| 15. | Stability of differential, REF of transformer/reactor | This test is performed to check the proper operation of differential and REF protection of transformer and reactor by simulating actual conditions. Any problem in CT connection, wrong cabling, or relay setting can be detected by this test |
| 16. | Frequency response Analysis (FRA) measurement | Before commissioning, this test is required to make sure the transformer's active part has not suffered any severe impact/twitch during transport of transformer |
| 17. | Winding resistance measurement | To verify any broken strings and high contact resistance, loose connections in tapping change devices |
| 18. | Vibration test (for reactors only) | To check any abnormal vibration of shunt reactor during operation |
| 19. | Dissolved gas analysis (DGA) of oil sample | Oil sample for DGA to be drawn from transformer main tank before commissioning for base data and after 24 h of charging subsequently to ensure no fault gas developed after first charging. DGA of oil sample helps in concluding the generation of gas and materials involved in the urgency of righteous measures to be taken by the operator |

the top, proper tightening of the breather to the connecting pipe and condition of gasket is to be ensured.

**Visual Check for Conservator's Oil Level**: The transformer conservator consists of a magnetic oil-level gauge (MOG); each transformer conservator of the main tank is fitted with one MOG to indicate when the oil inside the conservator tank reaches minimum or maximum. The MOG dial has a scale from empty to full; it also has some intermediate divisions such as 1/4, 1/1, and 3/4. During full load condition, the temperature of the insulating oil increases accordingly, insulating expanded oil and the oil level goes up. After the full load, when the transformer load decreases the oil temperature reduces in turn reducing the volume of oil. But it is very essential to maintain the minimum oil level in the transformer conservator tank even at the lowest possible temperature. The entire transformer top of the conservator tank is fitted with a MOG. A lightweight hollow ball is floated on the insulation oil inside the conservator tank. The float arm is attached with bevel gear and the float goes up and down depending upon the oil level and consequently the float arm alignment changes. The MOG is float operated through the tank wall with the magnetically coupled float mechanism to the dial indicator. The float rotates a magnet inside the tank as the level increases beyond a certain level with the increase in level. The oil level indicated by the MOG and top oil temperature should be observed. The top oil temperature reading should be used to correct the oil-level gauge reading. The corrected oil level should be in the position corresponding to normal as mentioned in the OEM instruction manual (35 °C) range. If the level of oil is standard, there is no need to take extra action, but adding the oil and removal of oil from the conservator tank should be done if it is below or above the normal level, respectively.

**Checks on Temperature Gauges**: Oil temperature and hotspot temperature gauges are important for proper operation of the transformer. These gauges not only indicate temperature but also operate the fans and coolers by means of microswitches (mercury) that can be adjusted for various temperature settings. These gauges should be calibrated on a regular basis onsite with portable devices or in the laboratory.

**Winding Temperature Indicator (WTI) Test**: WTI is used to indicate the temperature of the winding based on the manufacturer heat run test. It does not have the temperature-sensing device which is fixed in the winding hotspot; as it gives only an approximate value of the temperature of the hotspot winding, it should not be an accurate value. But it can be used to activate the cooling system as well as the top oil thermometer. Winding temperature thermometers are placed near the top of the tank; it is similar to topping the oil thermometer which has a separate thermometer well without the bulb [8].

A wire is wrapped between the thermometer and temperature-sensitive bulb or a heater coil can be inserted into the thermometer in same way. A current transformer (CT) as in certain transformers has three leads of windings and gives current in proportion to the winding current directly to the heater coil. In other transformers, the CT gives current to the heater coil via an autotransformer. The bulb is slightly heated by the heater and the temperature is shown at the indicating dial although the value shown is not the true temperature of the hottest spot. Based on measured temperature rises, or data from tests of a thermally duplicate transformer, bias current to the heating coil of the winding temperature indicator is factory adjusted to simulate the same gradient in degrees Celsius over the top oil rise as will be experienced by the hottest spot in the transformer windings. The manufacturer makes adjustments in these devices by altering the taps on the autotransformer or CT, or by altering the calibration resistors in the control cabinet.

WTI normally cannot be field tested or calibrated other than testing the thermometer. Current in the heater circuit is adjusted by the transformer manufacturer; the magnitude of this current should be known to allow calibration of the unit to be verified. The calibration resisters can be adjusted in the field if the calibration curves of the heater current versus hottest spot gradient for the transformer available with the manufacturer are provided to user.

### *5.3.4 Cooler Control Setting, Alarm, and Trip Test*

- Make sure transformer equipment is isolated, properly de-energized, and earthed.
- Temperature setting should be per standards. Only indicative values are furnished below. These values are to be taken seriously and confirmed with the instruction manual provided by the manufacturer.

- Use the local winding temperature indicator and move its pointer slowly to the first level of cooling value (say 65 °C).
- The cooler's fans which are set to the first level need to be checked in operation.
- Continue rotating the pointer to the second level cooling value (say 80 °C).
- Fans/pumps of those coolers which are set to the second stage need to be checked in operation.
- Continue rotating the pointer to the alarm preset value (say 100 °C).
- Check the alarm signal received in the control room is trustworthy.
- Set the tripping value (say 110 °C) by changing the indicating pointer.
- Whether trip signal has been received in the control room should be checked.

### *5.3.5 Calibration of the Winding Temperature Indicator*

- The winding temperature indicator bulb needs to be removed from the transformer pocket placed in the tank cover.
- The bulb is inserted into a calibrated temperature-controlled bath.
- The temperature of the bath is raised in 5 °C steps and the response of the winding temperature indicator verified in a 10-min cycle up to a maximum temperature of 130 °C. The tolerance for the temperature indication is ±3 °C.
- Now the temperature of the bath is lowered by 5 °C steps and the response of the winding temperature indicator in the 10-min cycle is checked. In parallel verify the transducer output; the allowable tolerance limit is ±3 °C.
- Verify alarm and trip switch configuration by changing pointer to set temperature. Use a multimeter to indicate these settings and record the operational values.
- After completing return the bulb to the pocket in the transformer cover. Care should be taken for matching levels between temperature indicator and pointer.

### *5.3.6 Secondary Injection Tests*

ONAN Rating

- Validate isolated cooler supply is in switched OFF state.
- Keeping constant temperature of 50 °C, change the winding temperature indicator bulb in the calibrated temperature-controlled bath.
- The rated current is injected into the winding terminal, a temperature indicator test kit is used for verification, and the result noted where the obtained result is same as the ONAN rating.

OFAF Rating

- A running pump is required to carry out these tests.
- Set the cooler isolator at ON position.
- Set the cooler switch to fans and pumps position.
- Ensure that pump is running.
- Turn on the heater coil and check the gradient for OFAF rating.
- Disconnect and set the cooler control switch to AUTO.

### 5.3.7 *Oil Temperature Indicator (OTI) Tests*

The thermometer consists of a cylindrical sensing body with a flange, capillary tube, and thermometer housing with dial and contact device. The measuring system is filled with a liquid that changes its volume at temperature variations and affects springy bellows [9]. The movements of the bellows are transferred to the pointer and signal contacts via a link system. The thermometer is provided with four microswitch-type signal contacts. The contacts can be set independently of each other. As to control and adjustment of the thermometer, see the instruction manual supplied by the manufacturer. This manual also gives settings recommended for the signal and start of the cooling equipment, respectively.

(a) Calibration of Top Oil Temperature Gauge

- Remove the OTI bulb and insert calibrated temperature-controlled oil bath from the pocket on the transformer lid.
- Increase the temperature of the oil from 0 to 120 °C step by step; increases in temperature should preferably be 5 °C or 10 °C intervals. Note the OTI, reading again the oil temperature range with tolerance ±3 °C.
- Note the oil temperature and instruct the alarm indicator (i.e., 95 °C) and the trip value pointer to 110 °C and check if they are operating using a multimeter across the said switches.

**Interpretation**: If the temperature difference between the reference temperature and dial indicator is more than 5 °C, it is suggested that it should be replaced with a healthy one. If it is impossible to replace the temperature gauge with a new one, or send the gauge for repair, put a temperature adjustment factor on the gauge to correct the error (addition or subtraction to the dial reading) to get the correct temperature reading.

Once the alarm lower turning on the settings of pump by the same correction factor mentioned above. These are pressurized systems and if out of calibration the indicator will read low. During field testing of the RLA it was observed that some of these gauges read 10–15 °C lower than normal temperature. The transformer gets in dangerous conditions as it allows continuous running in hotter than intended temperatures due to late activation of alarms and cooling. Transformer life may be shortened to failure if the thermometers are not properly tested and errors are corrected.

**Checks on Gas Pressure Relay**

There are two types of gas pressure relays commonly used in the transformer/reactor. The most common type is mounted under the oil and a similar type is mounted in the gas space. Internal arcing in liquid-filled electrical power equipment produces extreme gas pressure; it can severely damage the transformer and also pose severe danger to line personnel. The gas pressure relay is used to reduce the degree of damage by quickly activating the transformer protection system. The manufacturer's recommendations should be referred to for adjustment, repair, or replacement of an improperly operating device.

**Pressure Relief Devices (PRDs)**: After the pressure release, the PRDs are spring-loaded valves that automatically reclose. When the PRD does not operate properly, the pressure inside the transformer is not relieved adequately. As a result, within a few seconds catastrophic rupturing of the tank occurs, spreading flaming oil over a wide area. Once a year, or as soon as possible after the internal fault, inspect PRDs and test the working condition. Switch the alarm circuits ON and make sure whether the correct communication point is activated.

**Note**: During operation if the PRD operates, then do not try to re-energize the transformer. All LV tests may be required to be carried out before re-energization and the oil sample sent for testing at DGA.

**Sudden Pressure Relays (SPRs)**: A sudden pressure increase by arcing can be detected by SPR. The operating time is very fast and designed to operate before the PRD operation. Pressure is exerted at the bottom of the transformer tank; when a sudden excessive pressure develops this would move the spring operating pin. This operates like a switch generating further alarms and a breaker trip signal is generated. Once the relay responds successfully, the relay cap must be detached and the switch reset to normal by resetting the button.

**Note**: If this relay operates, then do not try re-energizing the transformer until the exact cause of operation is determined and corrective action taken.

## *5.3.8 Checks on Buchholz Relay*

(a) **Alarm Checks**

- Connect the tubes of the "bomb" to the copper pipe from the relay test cock (top of the relay) and foot pump.
- Charge the Buchholz pressure bottle with the foot pump to a pressure of 4 bar (60 psi). Measure the volume of gas trapped inside the window and mark.

- Keep the valve on the tank side closed. Open the relay test cock. Alternatively open the drain valve/plug of the relay.
- Very slowly release the valve of the Buchholz pressure bottle to allow air into the relay.
- Check the control room to ensure that the gas alarm has indicated once a sufficient amount of air has entered the relay.
- Check for correct operation using a multimeter (continuity of alarm contacts) connected across the alarm contacts in the relay.
- Close the relay test cock. Close the valve on the Buchholz pressure bottle.
- Vent all air from the gas and oil relay; close when a steady stream of oil is discharged into the bucket.

(b) **Trip Check**

- Charge the Buchholz pressure bottle to a pressure of 2 bar (30 psi).
- Keep the valve on the tank side closed.
- Open the relay test cock.
- Open the drain valve/plug of the relay.
- Quickly open the valve on the Buchholz pressure bottle to allow a full surge of air to enter the relay.
- Check the control room to ensure the surge trip relay flag has operated.
- Check correct operation of relay trip switch contacts using a multimeter (continuity of trip contacts) connected across the trip terminals.

(c) **Gas and Oil Relay Surge at Pump Activation**

Check the status and conditions of the alarm and trip contacts of the Buchholz relay as soon as the oil pump starts. This check is carried out manually and automatically so that any spurious alarms and trips do not result in a faulty condition.

(d) **Results**

If the relay operates correctly close the bomb valve, disconnect from copper pipe work, and vent all air from the relay petcock. If the relay fails to operate repeat the alarm check at successive incremental pressures of 0.25 bar (3.5 psi) up to 4 bar (60 psi) until the relay operates. Operating pressure is to be recorded.

If the relay fails to operate, isolate the relay from the expansion vessel and the tank using the valve (see the concerned valve on the general arrangement drawing). Drain the oil from the relay mechanism.

Blank off each open flange while investigating the relay fault to ensure that no contamination enters the system. If the switches in the relay are faulty replace them. Alternatively, if the relay mechanism is faulty replace that. Refit the relay into the pipe work using new gasket material. Vent all air from the relay. Open the relay test petcock and vent all air from the system after completion of test.

### 5.3.9 Checks on Cooling System

Large power transformers are fitted with some type of cooling system. Cooling systems generally consist of combinations of radiators, pumps, and fans. The cooling surfaces should be inspected regularly and cleaned of foreign particulates. This is especially important in the case of fan cooling. The cleaning should be performed with high-pressure water flushing and proper safety measures such as by covering the fan motor, so that the high-pressure water does not enter it [10]. During flushing, it is likely that the radiator fins get bent. They should be straightened immediately after cleaning. Finally dry cloth and cleaning solutions are used in the cleaning process. Normally, no measures are necessary for keeping the internal cooling surfaces clean as long as the oil is in good condition. If, however, sludge formation has set in, the sludge may deposit on horizontal surfaces in radiators and coolers. If that occurs, the radiators and coolers should be internally flushed by new clean oil in connection with the oil exchange.

The cooler bank should be inspected regularly. Cleaning of coolers is carried out by pulling off the tube packets and making them accessible for cleaning. This activity is to be carried out in consultation with the manufacturer. Generally all the fan motors have permanent lubrication in their bearings and also have dual sealing rings. The motor bearings are axially clamped with spring washers. If the noise level of the fan increases, all mounting supports should first be retightened.

### 5.3.10 Cooling System—Fans

(a) Visual inspection should be done when the transfer is onload to ensure fans are operated at the rated speed and their airways are not barren, thus saving the blades from damage. Any fan that is running at less than its design speed will be obvious to the naked eye. For more precise measurements, a tachometer may be used.

(b) The rotation of the fan blades should be observed to ensure that the air flow is in the proper direction for the type of device involved. The observation may be facilitated if it is performed at a lower than normal speed either during startup or immediately after switching off. Corrections to rotation should be made as indicated by inspection.

(c) All fans that are not running at design speed should be replaced. After stopping fans, any obstructions to air flow should be removed and any damaged fan guards or blades should be replaced or repaired.

(d) If the motor is fitted with heaters they should be switched on whenever the motor is standing. Check for buildup of moisture in the motor. Drain holes are provided at the lowest part of the motor and are fitted with plugs on totally enclosed motors (wherever applicable). These plugs should be removed to allow any moisture to drain away.

(e) Regular greasing of the bearings should be carried out and as a general guide, one or two shots from a grease gun should be sufficient at intervals of 1000 running hours. It will only be necessary to maintain the grease in the bearings.
(f) It is recommended that at least every two years the bearings and housings be flushed out in white spirit and inspected for wear. Worn parts should be replaced where necessary and repacked with grease.

**Cooling Fan Controls**: The cooling fan manual and automatic control system can be operated; however, automatic operation is associated with the load and energization.

**Manual Control**: This turns ON for a brief period to ensure that each stage has sufficient voltage to operate. Fan operation should be observed and the oil pumps' operation and their flow gauges checked. Refer to the manufacturers' recommendations for any malfunctions.

**Temperature Control**: Remove the temperature bulb from its well on the top of the transformer; put the master control in automatic position. Using a calibration instrument slowly raise the temperature of the bulb and observe for proper calibration/operation.

**Load Control**: Verify the CT controlling secondary current operation. Remove the secondary lead after shorting out the CT secondary (if the transformer is energized); further current should then be injected in the control circuit and its level varied to observe proper operation.

**Caution**: Extreme care should be observed when performing operations with the secondary of an energized CT; otherwise if the CT secondary is left open by mistake, catastrophic results may occur without warning.

### 5.3.11 Cooling System—Pumps

(a) Cooling pumps are generally centrifugal-type pumps and as such will pump some oil regardless of their direction. The pump has a direction and flow gauge/indicator fitted on the top of the pump. The pumps should be manually turned ON and OFF and the action of each pump's flow gauge should be observed while the pump is coming ON. The oil flow should cease or be at a bare minimum before energizing the circuit.
(b) If the movement of the flow gauge flag is sluggish where 3-ph motors are used, there is an indication of reverse rotation. Reverse any two electrical leads supplying the suspect pump and re-energize. The movement of the flow gauge flag should now be more prompt.
(c) To determine excessive bearing wear remove the pump for visual inspection. After removing the pump from the system, the end play of the shaft should be measured. The impeller and impeller housing should be examined for any wear. The manufacturer's guide should be consulted to determine if excessive bearing wear exists as indicated by the amount of shaft end-play observed. Any

indication of wear on the impeller and impeller housing is indicative of excessive thrust-bearing wear.

(d) Partially shorted motor windings and other electrical problems with the pump motor will cause generation of combustible gases in the transformer oil DGA because oil directly flows through the pump motor during normal operation. Routinely or after detecting an abnormal level of combustible gas in the transformer oil DGA, the current flowing in each electrical terminal of each pump motor should be accurately measured at the running condition of the pump.

(e) The oil flow gauge shows only whether there is oil flowing through the pump. It is not indicative of the velocity of the oil or the condition of the pump. After making sure that the oil cooling pumps are ON, the flow gauge should be observed for indication of flow. If the pump is ON and no flow is indicated, the sensing unit may be defective. If the pump is turned OFF and the flow gauge continues to indicate, the gauge is probably stuck in the flow position and the sending unit or the entire gauge may need to be replaced. It is normal for a gauge to continue to indicate flow for a brief period before indicating OFF (no flow), following the turning off of a pump. This is due to conservation of momentum with respect to oil.

### *5.3.12 Operational Checks on OLTC*

**Tap Changer Handle Operation**: Move the tap changer handle up as well as down over its entire range; check the operation before trying to attempt electrical operation. Also keep in mind that while the handle is inserted, the interlock switch of the said handle will not allow electrical operation. In addition to this, check tap positions for the single-phase tap changers and make sure that they reached together at the motor drive unit head. Continuity and discontinuity of the tap-changing operation is verified by connecting a multimeter across the high- and low-voltage sides and changing the position from minimum to maximum.

**Maintaining Circuit**: The correct sequence of the maintaining circuit needs to be checked by hand, winding the units to halfway through a tap and then removing the handle. Power up the drive motor and also make sure that the motor runs to drive the tap changer handle in the same direction.

**Limit Switches**: Limit switches' operation needs to be thoroughly checked at both ends.

**Drive Motor**: Put tap changer in the middle position and verify the rotational direction. Measure the starting and running currents in both the higher and lower modes of operation and also measure and record these values. Set the motor's overload to 10% above normal running current.

**Raise and Lower Control**: (1) Step-by-step relay operation; (2) move tap changer one tap at a time.

**Out of Step Relay**: Shift the first tap changer in a three-phase bank to be one position out of step with respect to the other two phases. Check the alarm's condition to ensure it is activated properly. In the same way repeat this process for the other two tap changers.

- First check the suitable operation of the local/remote switch.
- Using local control, raise and lower the tap changer over its full operational range.
- Repeat the above using the remote raising and lowering facility over the full range.
- Check that neither the local nor remote switches can operate the tap changer when it is in the other mode and vice versa.
- Hold the raise and lower pushbuttons in following a tap change to ensure it only moves one tap at a time hence checking the step-by-step relay.

**Tap Change Position Indicator**: Check the mechanical position indicator to ensure accurate indication in all positions. Also check that the remote instrument operates in the same sequence to the same accuracy. With the remote instrument unconnected, operate the tap changer and measure the output from each binary coded matrix.

**Tap Change Incomplete Alarm**: Check whether the operation of the tap changer is incomplete by keeping the winding unit on hand halfway through a tap changing, including alarm working, flag relay, and also monitoring if their operation is right as well as time of operation.

**Operation Counter**: Ensure that the tap changer counter operates and reads correctly.

**Remote Indication**: Check the remote indication switch and control facility to prove that the output terminals of the said arranging kiosk are working fine.

**Tap Changer Protective Relay**: Verify the tripping characteristics of the relay by opening the cover and pressing the "Trip" button to check the operation of the transformer CB; ensure proper working function and then close the cover. The following checkpoints/guidelines for inspection and maintenance should be addressed and the manufacturer's service engineer should be consulted for details of maintenance/overhauling activity to ensure the absence of problems and ensure proper operation in the future.

The following checks are to be carried out on OLTC.

(a) Manual operation on all taps and handle interlock switch.
(b) Local and remote operation and L/R switch.
(c) Matching of tap changer positions and tap position indicator.
(d) Check the working order of the tap selector handle; exchange the selector and arcing transfer switches.
(e) Drive operations.
(f) Counter, position indicator, and its co-ordination with mechanism operations.
(g) Limit switch operation.

(h) Mechanical blocks integration.
(i) Proper movement of the hand-crank and its interlock switch.
(j) Physical wear and tear of tap selector.
(k) Free movement of the external shaft assembly.
(l) Extent of arc erosion on motionless and variable arcing contacts.
(m) Inspect the end boards for tracking and cracking the faults.
(n) After filling with oil, manually crank during entire range.
(o) Oil BDV and moisture content (PPM) to be recorded and measured to a minimum BDV of 40 kV and moisture content of 10 PPM.

### 5.3.13 Checks on Bushings

After one month of service and on a yearly basis, make sure that the porcelain bushings are free of dust and dirt; otherwise proper cleaning is required. Particularly in the areas where there is high air pollution, salt regions, cement dust, or chemical substances, cleaning is required regularly at short intervals. If the bushing is damaged/highly polluted, it causes excessive leakage current and a visible mark appears as treeing (carbon tracking) on the surface of the bushing and may cause flashovers if the bushings are not cleaned and replaced properly.

By looking in the oil-sight glass, check the oil level in the bushings because the level does not change with varying temperature. Stuck oil gauge pointers coherent with an oil leak can cause a collapse in the bushing and possible damage in the transformer and switchyard equipment. If the oil level is low, and the external leakage not visible, there might be internal leakage around the lower seal of its tank. In that case refill the bushing with the same oil as per the standard. An old bushing is replaced when it has more than 25% of bushing oil capacity; the bushing is sent for repair or it may be completely scrapped, if unrepairable at the site.

Inspect the bushing using an infrared camera regularly after one month of service and on a yearly basis. In the case where any of the phases shows an eccentrically high temperature, there is a possibility of poor connection; line personnel have to verify the top connection first, however, a poor connection inside the transformer tank leads to higher temperatures at the top also. Furthermore these faulty connections should be detected by finding the hot metal gases such as the ethane and ethylene ratio in the DGA.

**Caution**: When the temperature is high don't open the oil fill plug of any bushing because pressurized hot oil gush may result in a burn injury. Generally a safe temperature range is between 15 and 35 °C. Some bushings may have gases such as nitrogen in the form of a blanket on top of the oil layer which pressurizes it as the oil expands due to temperature. Check with the manufacturer while topping up oil or opening the bushing. Also see the directions in the special information documents about bushings that are included in the transformer instruction manuals at delivery.

## 5.4 Connectors

In order to avoid extreme temperature rises at the electrical connections of the transformer, check screw joints and retighten the screws according to the manufacturer's manual. For transformers in service, a thermovision camera may be used.

## 5.5 Maintenance of Insulating Oil

Mineral oil is used as an insulating fluid in transformers/reactors. In addition to acting as an insulating fluid, it also acts as a heat transfer medium to carry off excess heat generated by the losses of power transformers. Tests cover the determination of certain quantities, primarily degradation constituents in in-service transformers and the diagnosis of these results wrt the power transformer condition [11, 12].

A sampling technique should ensure that the specimen taken is representative of the insulating oil contained within the equipment. Natural contaminants exist within the body of sampling valves; therefore valves should be flushed before extraction is performed in order to ensure that sample integrity is maintained. A sufficiently large sample should be withdrawn to cover all tests as listed below. Typically 1 L oil is enough for carrying out the following tests to ascertain continued serviceability of any insulating oil filled in the transformer/shunt reactor or any other electrical apparatus. The tests explained above are carried out at specific time intervals. However, for carrying out DGA, syringe sampling is recommended.

- Dielectric strength sampling
- Water content sampling
- Specific resistance sampling
- Dielectric dissipation factor
- Flashpoint sampling and recoding
- Total acidity sampling
- Interfacial tension sampling
- Sludge content (for aged oils only)

The following tests are generally performed to determine the various properties of oil meeting certain conditions.

- Kinematic viscosity
- Pour point sampling
- Carbon type composition
- Oxidative ageing
- PCB content and density
- Oxidation stability and SK value

The diagnosis of the equipment containing oil is evaluated with the following tests; they do not check the quality of the oil.

- Dissolved gas analysis (DGA)
- Furfur-aldehyde analysis (FFA)

Give special attention to diagnostic tests, for the oil-filled equipment, and not the oil itself. The DGA evaluates the composition of gases dissolved in transformer oil and estimates most probable incipient faults in the transformer using various interpretation techniques per standards such as IS9434, IEC 60599, IEC 60567, and IEEE C57.104. The furfur-aldehyde analysis is generally for checking the health of the paper insulation in the windings by measuring the 2-furfur-aldehyde released in the oil when the insulation paper degrades. It is to be understood that these two tests are diagnostic and trend related and not analyzed on absolute values of results. Normally insulation oil is checked against dielectric and moisture content in an onsite test to determine the transformer oil condition.

## 5.5.1 Tests for Dielectric Strength (BDV)

To do the BDV test, a 2.5-mm gap is maintained between the electrodes; a further six tests are performed on the oil to check its breakdown voltage. The six-digit value in the result is taken into account to check the acceptance condition for 70 kV. Refer to IS-6792 or IEC-60156 for the detailed testing procedure.

**Tests for Moisture Content (Measured in ppm)**: Inject a sample of the oil in an automatic moisture content test kit. Depending on the type of test kit, the moisture may be measured by µg $H_2O$ and the measured figure could be divided by weight of the injected oil sample in grams, thus giving parts per million units. Generally moisture content would be maintained less than 10 ppm when the transformer is in service condition [13, 14]. As this test is highly operator- and kit-maintenance dependent, it is advisable to send the sample to an authorized laboratory only.

**Note**: DGA frequency for new equipment under commissioning should be just before charging, after 24 h of charging, 7 days, 15 days, one month, and three months after charging; thereafter six monthly DGA to be followed. The frequency of sampling can be changed depending on test results.

**Frequency Response Analysis (FRA)**: The FRA test is conducted on transformers and reactors to determine the frequency response of windings. The reference frequency responses obtained during laboratory testing serve as "fingerprints" to monitor the condition of the transformer or reactor during service.

The frequency response of an electrical winding is obtained by application of sweep frequency (sinusoidal). The winding will have a characteristic frequency response for the applied signal at different frequencies. The response is uniquely determined by the winding arrangement involved and any winding movement or other fault will modify the frequency response due to changes in inductance and capacitance. The sweep frequency voltage is applied through network analyzers. The frequency response of the winding is determined between the frequency ranges of 10 Hz to 2 MHz.

The FRA test is performed on one winding of the electrical equipment at a time. The transformer/reactor should be electrically isolated from any other electrical connections or systems, including earth connections during the FRA test. The two end terminals of each winding should be made available for measuring the frequency response across the winding.

- For star-connected winding, the response should be measured across the terminal and neutral.
- For delta-connected winding, the response should be measured across two line terminals and in the case of open-delta, across individual winding.
- For autoconnected winding, the response of series and common windings should be measured separately.

For a transformer, it is normal practice to earth one end of every winding that is not being tested, leaving the other end open. Alternatively, all other windings may be left unconnected from each other and from earth. In every case, the termination of each winding for each test should be recorded. The frequency response of the winding is determined by plotting the ratio of the output from the winding to the input in the frequency range of 10 Hz to 2 MHz.

Alternatively frequency ranges specified by the customer can be selected. The test is normally conducted at maximum, mean, and minimum taps, in the case of windings having tapping. While making measurements at mean tap, care should taken to move the tap from higher voltage taps for proper comparison of FRA results of different phases of the same or different transformers.

The FRA results are analyzed for:

- Changes in response of the winding
- Significant difference between the FRA records of different phases of the same transformer
- Significant difference between same phase of identical transformers

The FRA test is primarily a condition assessment test and can be used in conjunction with other diagnostic tests for detailed analysis and interpretation of the transformer.

**Dielectric Frequency Domain Spectroscopy (FDS)**: In frequency domain spectroscopy (FDS), the frequency sweep measures the dissipation factor of the insulation components. The frequency versus tangent delta measurements method is called frequency domain spectroscopy. In this method, the frequency range is much enhanced, especially to low frequencies.

Figure 5.2 shows the S-shaped curve where the dissipation factor is plotted in the $y$-axis and the frequency is in the $x$-axis. The temperature and ageing curve move towards the higher frequency scale when the moisture content increases.

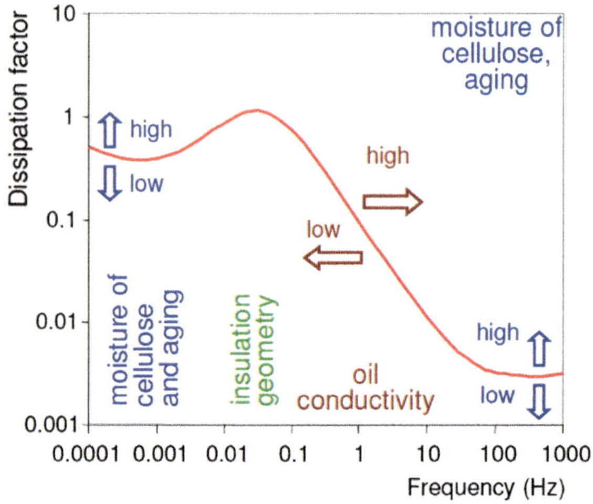


**Fig. 5.2** Dielectric frequency domain spectroscopy

**Table 5.8** Moisture content in paper

| S. No. | Insulation condition | Source: IEEE standard 62-1995 | |
|---|---|---|---|
| | | % Moisture by dry weight in paper (Wp) | % Saturation of water in oil |
| 1. | Dry (at commissioning) | 0.5–1.0 | <5 |
| 2. | Moderate to wet (lower number indicates fairly dry whereas large number indicates moderately wet insulation) | <2 | 6–20 |
| 3. | Wet | 2–4 | 21–30 |
| 4. | Extremely wet | >4.5 | >30 |

*Source* CIGRE DOC. No. 227. Life Management Technique for Power Transformer

Moisture content has greater influence in the lower and higher frequency ranges; the steep gradient shows oil conductivity in the midrange of the curve. The "hump" left side of the steep gradient curve data helps to calculate moisture content.

Table 5.8 may be considered for determining the level of wetness and taking necessary action for dryout.

**Partial Discharge Measurement**: PD techniques for detection and location are important for diagnostics, as they not only help to identify the inception of damage caused, but also assist in monitoring the evolving and deteriorating situation affected by the various stress factors existing in the service condition. There can be various reasons for PD inception. It can be the result of electrical stress caused by mechanical deformation, overheating of an insulated conductor, or can even be the result of inherent defect. The PD detection at the site can be done by various techniques such as

- Dissolved gas analysis
- Conventional IEC 60270
- Acoustic measurement
- RFI measurement
- UHF measurement

All these techniques have their strengths and weaknesses in terms of sensitivity and accuracy. The sensitivity and effectiveness of these techniques depends upon the type of defect and location of the fault.

## References

1. Franklin AC, Franklin DP A practical technology of the power transformer, 11th edn. J&P Books
2. Transformers. Bharat Heavy Electricals Limited, Bhopal. Tata McGraw-Hill Publishing Company Limited, New Delhi
3. Rabins L (1956) Transformer reactance calculations with digital computers, communication and electricals. 75:261
4. Sealey WC Equivalent ambient temperature for loading transformers. Allis Chalmers (USA)
5. Indian Standards Related to Power Transformers:(a) Specifications: i. IS 1885 (Part XXXVIII)-1977 Electrotechnical vocabulary, part XXXVIII Transformers (first revision); ii. IS 2026 (Part IV)-1977 Power Transformers: Part IV Terminal marking, tapping and connections (first revision); iii. IS: 11171-1985 Dry Type Power Transformers, (b) Selection, Installation, Operation and Maintenance: i. IS: 1866-1967 Code of practice for maintenance and supervision of insulating oil in service; ii. IS: 1886-1967 Code for practicing for insulation and maintenance of Transformers; iii. IS: 5216-1969 Guide for safety procedures and practice in electrical work; iv. IS: 5728-1970 Guide for short-circuit calculations; v. IS: 6132 Shackles, Part-1: 1971 General Requirements, Part-2: 1972 Dimensions of D-Shackles; vi. IS: 6600-1972 Guider for Over-Loading of Power Transformers; vi. IS: 8478-1977 Application Guide for On-Load tapchangers, (c) Transformer Oil and Oil Testing: i. IS: 335-1983- New insulating oils for transformers and switchgears; ii. IS: 1866-1978-Code for practice for maintenance and supervision of insulating oil in service; iii. IS: 6792-1972 Method for determination of electric strength insulating oils, (d) High Voltage Testing: i. IS: 8690-1977 Application guide for measuring devices for high voltage testing; ii. IS: 2165-1977 Insulation coordination, (e) IEC Publications: i. 76-1 (1976) "Power Transformers"; ii. 214 (1976) "on-Load tapchangers"; iii. 606 (1978) "Application guide for power transformers"
6. Transformer engineering, A treatise on theory, operation and application of transformers by GE (USA)
7. Transformers for electric power industry by Westinghouse (USA)
8. IS: 2026: Specification for power transformers
9. IS: 6600: Guide for loading of power transformers
10. IEC loading guide for oil immersed transformers. IEC publication 354

11. Reynolds PH (1975) Partial discharge measuring techniques. In: International HV symposium, Zurich, September 1975
12. Condition Monitoring, IPG International Generation, January 1993, pp 33–42
13. Studies of insulator pollution under industrial contaminated conditions. Technical report, CPRI, Bengaluru
14. Proceeding of seminar on "Distribution transformer failures", CBIP, New Delhi, February 1997, pp 1–36

# Chapter 6
# International Best Practices in Operation and Maintenance (Advanced Gadgets)

## 6.1 Introduction

Aiming to track every unit of energy injected into the system, power utilities are increasingly being attracted towards reliable and smart technology driven by distribution system tools and tackles [1]. These devices came into existence with a promise to improve reliability of the power system by identifying the difficulties in operation and maintenance in power distribution systems. Using such equipment will improve the power supply quality along with reliability. Hence technical loss will be reduced and the system can be stabilized as a whole.

To increase the efficiency of the staff who are already working in DISCOMs improves building operational practices, and reduces utility costs; it is a necessary and emerging trend to create training/awareness programs on best practices in operation and maintenance (O&M) programs. The government of India facilitating these programs with the help of nodal agencies including MoP, PFC, NPTI, and IITs, empaneled power training institutes (PTI) and resource institutes (RI) for the development of material and the like. Power utilities worldwide have adopted the use of new techniques in tools and equipment for precise measurement of required parameters and timely saving of manpower. Failure of equipment such as transformers and circuit breakers incurs considerable cost and to avoid such failure the latest technology is incorporated in tools and equipment. The implementation of best practice in an O&M program can reduce energy usage by 5–20% without any substantial capital investment.

## 6.2 Reclosures/Auto Reclosures

In a transmission overhead line 80–90% of the faults are transient in nature; they most commonly occur in overhead lines due to lightning or temporary contact with a foreign object. Normally the distribution feeders are radial feeders protected by



J. Raja et al., *Practices in Power System Management in India*,
Power Systems, https://doi.org/10.1007/978-981-10-2972-1_6

timelag over current relays with instantaneous and delayed tripping functions. The circuit breaker is equipped with a switch mechanism that can automatically close the breaker after the occurrence of a fault and is called a recloser or auto closer.

It is located on the distribution feeder network (as shown in Fig. 6.1) to detect and interrupt transient/momentary faults. These reclosures of suitable size (i.e., load current and fault current) operate instantaneously to clear the fault followed by one or two reclosing shots with predetermined delays. In the case of a permanent fault, the reclosure remains tripped off until manually commanded. The result is increased availability of supply. It has faster single- phase reclosing and automated loop capabilities and is mainly used in radial feeders. The reclosing structure depends upon the dead time, reclaim time, and number of shots. The dead time setting mainly depends on system stability, load, circuit breaker characteristics, fault path ionization time, and protection reset time.

A recloser is designed in such a way that when the circuit breaker is reclosed on a permanent fault, the recovery time is sufficient for the protection relay's operation. Inverse definite minimum time (IDMT) or definite time over current and earth fault relays and spring winding are most commonly used to protect high voltage (HV) lines. The operating time can vary from 10 to 30 s, depending upon fault level. In the case of definite time protection the setting can be 3–10 s. It is common practice to keep a setting of 30 s on a HV autoreclosure scheme [2].

One or two shots are generally used in a distribution network. It depends on the frequency of the transient faults and the system conditions. However, three or more shots do not imply any significant improvement but substantially increase the stress on the breaker and other network apparatus.

**Fig. 6.1** Installation of reclosure

## 6.3 Sectionalizer

The sectionalizer (Fig. 6.2) is an efficient and effective electronic device and is very useful for an outdoor medium-voltage distribution network in combination with automatic circuit reclosure protection (Fig. 6.3). The function of this device is to count the number of fault occur on the medium-voltage line and open it active when the line is de-energized. The interrupting device is either upstream reclosure or a circuit breaker in the section. Because the fault currents are high, it is only used to save the fuse. Therefore for this device the current rating or time current rating in co-ordination with load- or sourceside fuses is absolutely not required.

If the fault was not cleared by a loadside protective device such as fuses or reclosures, then the sectionalizer must control the fault. The electronic device will actuate when the current in excess of the threshold is typically 300 mA or more, but as soon as the actuating current reduces, it stops counting/sensing the load current, assuming that the loadside fuse or reclosure has cleared the fault. In the event of an actuating current, if the electronic sectionalizer does not sense a 300-mA current, it will assume that the sourceside reclosing device cleared the fault, and the sectionalizer will increase the count. These selections are not co-ordinated with the associated source of the loadside protection device and have no direct communication. The above- mentioned conditions are based on the current measurements at the sectionalizer locations.

A typical installation on a double-pole structure is shown in Figs. 6.2 and 6.3. The sectionalizer is a SF6 insulated outdoor pole-mounted double- break family for overhead lines and designed for a modern remote-controlled distribution system. It offers maintenance-free operation in climatic conditions such as a salt-laden atmosphere, corrosive industrial pollution, and snow and ice. The device can be operated by electrical or remote control. For electrical control, a motor device and electrical control cabinet are required and for remote control, communication protocols and communication media are required.

**Fig. 6.2** Typical sectionalizer

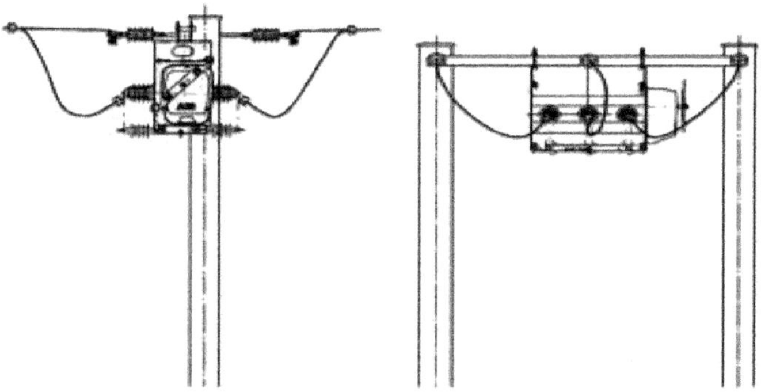

**Fig. 6.3** Installation of sectionalizer in outdoor medium-voltage line

## 6.4 Amorphous Core Transformer

Amorphous core distribution transformers of 11 kV and up to 500 kVA are available in India. Table 6.1 highlights the comparison of the losses of conventional and amorphous distribution transformers [3].

On the basis of technocommercial comparisons, utilities have initiated the separate procurement of amorphous core transformers. Although post procurement, repairing and maintenance have been the challenge, for evaluations of performance utilities need to compare the performance based on the total owning cost (TOC) for the predefined period.

## 6.5 Dry Type Transformer

The trend of using noninflammatory transformers is increasing day by day on account of high land cost, environment-friendly requirements, and stringent safety norms for public works and utilities such as big residential and business buildings, commercial malls, industries, medical and sports facilities, museums, theaters, art galleries, special industries such as mining, chemical oil and gas industry transportation industries such as ships, metro systems/trains, tunnels, and so on.

**Table 6.1** No-load loss of conventional and amorphous transformers

| Rating | Conventional transformer | | Amorphous transformers | |
|---|---|---|---|---|
| kVA | No load loss (Watt) | Full load loss (Watt) | No load loss (Watt) | Full load loss (Watt) |
| 25 | 100 | 685 | 68 | 466 |
| 63 | 180 | 1235 | 123 | 844 |
| 100 | 260 | 1760 | 176 | 1192 |

An O&M engineer should have thorough knowledge regarding the maintenance of these transformers; it is important and should be top priority. The maintenance of the said transformers is similar to oil-filled transformers but requires separate coverage. These are classified as ventilated, nonventilated, and sealed unit transformers (Fig. 6.4). The key aspects of maintenance for each of the above transformer types are essential considering the significant differences among the three.

- A ventilated dry-type transformer is designed for indoor and outdoor applications. It is just like an oil-filled transformer; the air or gas serves as an insulating medium and also dissipates the heat from the windings.
- A nonventilated transformer is totally closed and it is suitable where the atmosphere contains conductive, corrosive, or combustible material that might damage the transformer or lint and dust which might block the ventilation passage.
- A sealed transformer is like a nonventilated transformer in preventing entry of any surrounding air. An inert gas is filled at positive pressure in the said transformers.

Similar to oil-filled transformers, dry-type transformers require routine inspections and periodic checks. The periodic inspections depend on transformer capacity, classification, environment, and load conditions. Nonsealed transformers especially require more periodical checkups; because the transformer is in contact with atmospheric air, any dust particle or vapors it carries can contaminate the internal workings and increase electrical stress on the components. Thus the ventilated transformer requires more periodic maintenance than sealed transformers.

## *6.5.1 Maintenance of Dry-Type Transformer*

The greatest advantages of this type of transformer are that they require less maintenance; nevertheless, it is necessary to ensure dust does not accumulate, as it

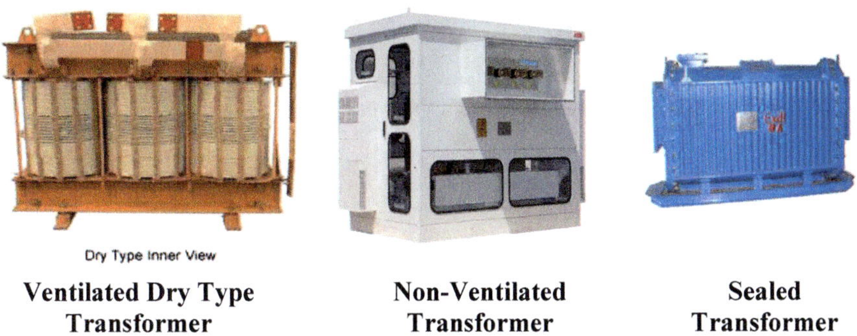


**Fig. 6.4** Various advanced kinds of transformers

causes a loss of cooling efficiency and hence subsequent loss of power; there should be inspection for any connections and structure deformation.

(a) **Routine checks and resultant maintenance**

A transformer operational record should be separately maintained. A complete checklist with essential observations and readings should be developed for each transformer and properly maintained in the records in the event of an extraordinary occurrence that could affect the performance of the related transformer and even every event related or not with the operation of the electric system. It is recommended to monitor the transformer with daily or weekly readings of the temperature, load, and voltages of the transformer. Ventilated and nonventilated dry-type transformers have indicating devices including temperature, liquid level, and pressure. Thus, routine checks are subject to human observation as they involve visual and auditory observations. Also sealed dry-type transformers have pressure gauges in them and they should also be routinely checked.

### 6.5.2 *Visual Inspections*

Visual inspections should cover:

(i) Due to poor contacts, overheating occurs at the HV- and LV-side terminals. If that happens, the O&M engineer should clean the areas of contact and tighten the bolts and nuts.

(ii) Due to overload, insufficient calculation of cooling air and temperature of the cooling air above the predicted level causes overheating of the transformer. If that happens, the O&M engineer should reduce the load, increase the cooling, clean the cooling air chamber, check the circulating ducts, and ensure proper opening of the ducts.

(iii) When the voltage is higher than the predicted voltage and the transformer not evenly sealed, resonance with other surfaces around the equipment causes excessive noise from the transformers. If it happens, the O&M engineer should select the most suitable voltage and adjust it to the most suitable tap. The engineer also has to check loose metallic surfaces such as panels, closets, ducts, doors, and the like to reduce the noise. The resonance engineer has to install a flexible element between the terminals and the installed cables.

### 6.5.3 *Dust Accumulation*

One of the most critical factors for efficient operation of the said type of transformer is to keep it dust-free in order to prevent any degradation in its important features. If dust accumulation is excessive, first for safety reasons the transformer should de-energize and its side panels be removed. With the help of a dry cloth, duster, or vacuum cleaner remove all the dust accumulated on the transformer. Subsequently, compressed clean air can be used in the ventilation channels of the coil and, between the core and coil, the compressed air's pressure should be less than 5 atm. With the help of a cloth immersed in benzene, core, ironwork, and coil impurities can be removed, and the cleaning process repeated with a clean and dry cloth.

## 6.6 Ring Main Unit (RMU)

The ring main unit (RMU; Figs. 6.5 and 6.6) consists of medium-voltage switchgear and can house the fuse-switch disconnections. The medium-voltage switchgears generally include both the preconfigured and fully configured gas insulated solutions. While using the tested equipment a high level of personal safety should be ensured.

## 6.7 Packaged Unit Substation (PUS)

The packaged unit substation (PUS) is an enclosed construction; it avoids bird hits, human/animal contacts that may lead to fatal accidents, disruption of supply, and power loss as may happen in open execution. The PUS is built as a product in line with the thinking of the architects who are quite concerned with aesthetics, safety, and ease of installation [4].

The space to install a PUS is typically 10′ × 10′ which is approximately the space required for a car park. A saving of floor space to the tune of 80–85% can be achieved by installing unitised sub-station (USS) as compared to that required for conventional installation. A PUS can be installed with a tariff (TOD) meter. The current transformers (CTs) and potential transformers (PTs) for the TOD meter are housed in a separate cubicle with a tamperproof sealing arrangement (Fig. 6.7).

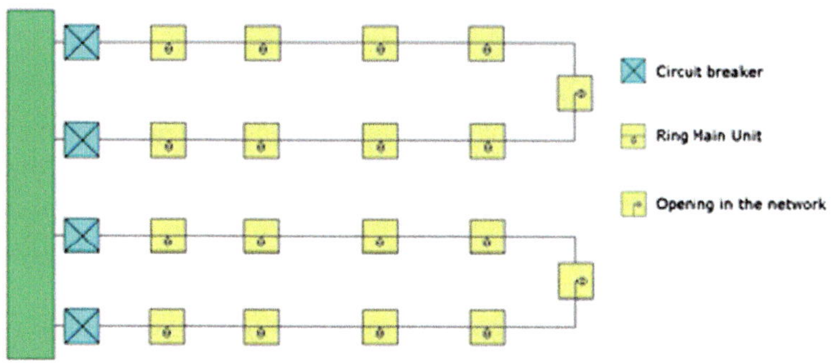


**Fig. 6.5** Typical 11 kV network and RMU locations

**Fig. 6.6** RMU unit and cable termination

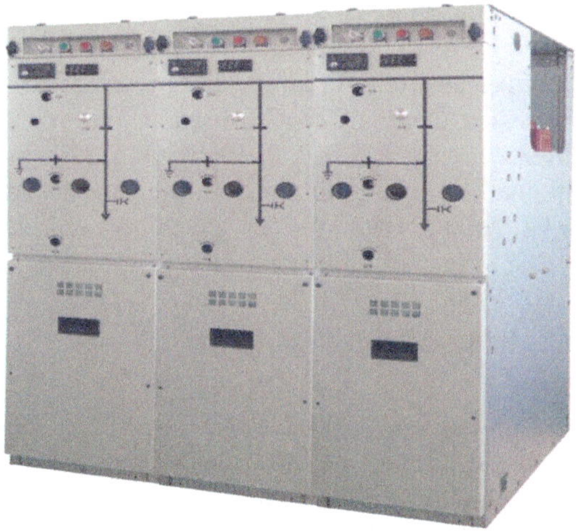

**(a) RMU**

**(b) Cable Termination**

A PUS mainly consists of the following equipment.

The PUS can be supplied via a preinstalled distribution-type transformer. The possibility of moving the substations depends on the type and the transformer size.

The PUS is supplied with internal cables/terminations:

- Medium-voltage cables from medium-voltage switchgear to the distribution transformer
- Low-voltage cables/bus bars from low-voltage switchboards to the distribution transformer

**Fig. 6.7** MV Switch gear

SF6 gas-insulated medium-voltage switchgear is connected using elbow connections. The transformer cable termination is through regular cable adapters or, if required, by the elbow connections. The necessary termination is provided in low-voltage cable/bus bars that are used between the distribution transformer and the switchboard. In accordance with standard norms and regulations, the PUS is designed to house various low-voltage solutions and individual requirements. The product range includes various types of fuse-switches and disconnectors, including molded-case circuit breakers. Other equipment can be supplied as per requirements.

## 6.8 Fault Passage Indicator (FPI)

Fault passage indicators (FPIs; Fig. 6.9) are used by distribution utilities. FPIs consist of current sensors with a light-emitting diode (LED). Indicators are strategically placed in the distribution network to identify a faulty network, as shown in Fig. 6.8. The FPI detects the fault when an appropriate threshold has been achieved [5]. The resulting sensor value output is periodically compared to the threshold. When an overcurrent situation occurs, a timing process is initiated to decide the activation of the indicator.

This system requires patrolling by the tripped main feeder or downstream CB on the faulted line; the said fault is mostly located between the last and the first nonflashing FPI/substation. FPIs are key in reducing the outage time on the network.

**Fig. 6.8** Fault indicator

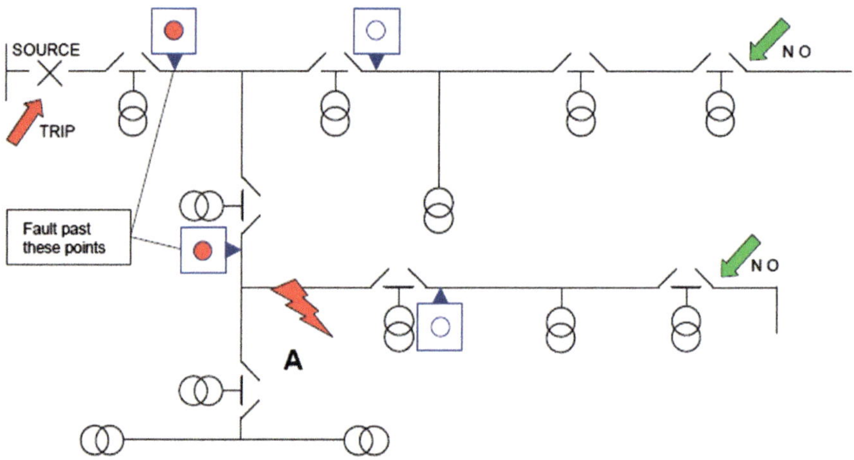


**Fig. 6.9** Indicators in distribution network

FPIs are divided into two types:

- Simple current criterion
- Simple directional criterion

## 6.9 Automatic Power Factor Controller

A power system has varying power demand on the supply system; accordingly the power factor also varies as a function of the load demands [6, 7]. It is very difficult to maintain a consistent power factor by the use of conventional methods (i.e., fixed capacitors and reactors). As a result, in overvoltage and saturation of transformers, penalties are incurred by electric supply authorities. The automatic power factor controller (Fig. 6.10) is a device that can automatically adjust the power factors without manual intervention and suit the power factor by load variation.

It is normally installed on the receiving facilities and power distribution at large consumers such apartment buildings and industries to enable full use of electrical infrastructure. It automatically detects reactive power requirements and accordingly controls the connection of power condensers to attain the ideal power factor. It has several capacitors grouped in several steps, and is also connected to suitable switching devices coupled with inrush current-limiting devices provided for each step. Normally the power factor sensed by the current transformer in the line side, and the KVAr required to achieve the target power factor are computed by the microprocessor-based automatic power factor controller relay.

## 6.10 Transformer Load Analyzer (TLA)

Sizing of the transformers is based on expected consumer load. The rating of the transformers is equal to greater than consumer load for a given period of time and the details of consumer particulars are taken from the DISCOM billing data. Figure 6.11 shows the data obtained from the analyzer.

To understand the better utilization of power distribution equipment, manpower, and materials efficiency, transformer load management plays a key role. The capital cost of the transformer can be viewed as a function of load capacity. Higher KVA capacity of the transformer is most suitable and economical to serve large consumers; similarly, an overloaded (or low ratings in connection with load capacity) transformer would be suitable for serving lesser loads to improve the power distribution equipment's life. The TLA has load analysis algorithms that model the thermal behavior and insulation ageing of the transformer, allowing the inherent overload ability of distribution transformers to be exploited.

To perform analyses such as transformer owning cost software (TOCS) and the TLA, much software is available. These types of tools can perform transformer

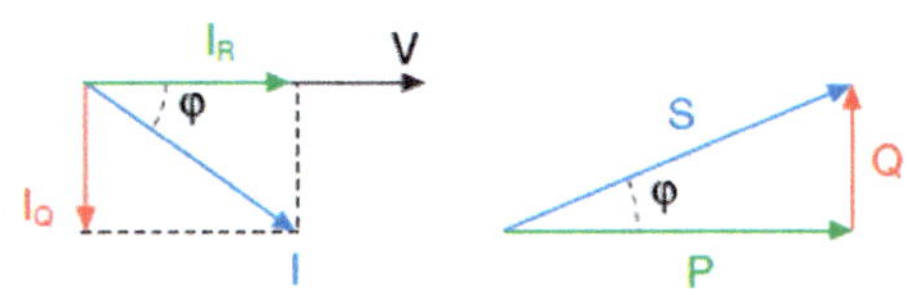


**Fig. 6.10** Vector diagram for power factor controller

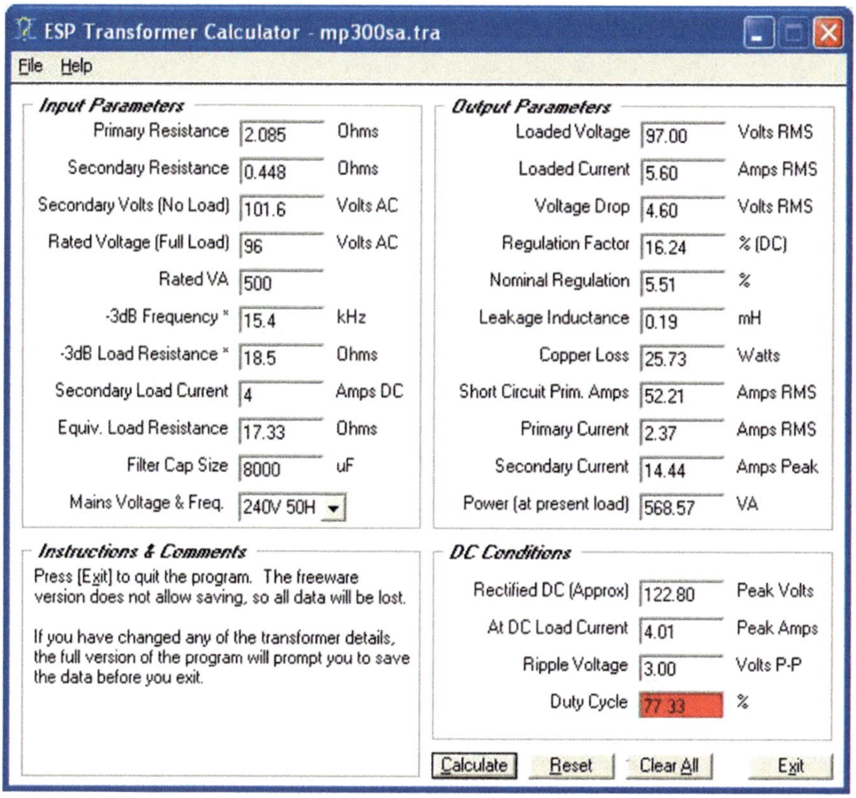


**Fig. 6.11** Screenshot of XFORMER

loading analysis on a seasonal basis, with hourly daily demand and ambient temperature curves representing the transformer behavior for all days in each of the four seasons in a year.

## 6.11 Advanced Tools for Operation and Maintenance

### *6.11.1 Thermovision Camera*

**Principle of Thermal Imaging**: All materials emit infrared energy when the temperature of the material is above 0° kelvin. The emitted infrared energy is then converted into an electrical signal with the help of an imaging sensor, then displayed on a monitor as a color or monochrome image [8]. Thermoscanner advantages and disadvantages are shown in Table 6.2.

**Table 6.2** Advantages and disadvantages of thermoscanner

| S. No. | Advantages | Disadvantages |
|---|---|---|
| 1 | It helps to prevent problems prior to their occurrence | The cameras cost a lot |
| 2 | It can be used to observe areas inaccessible or hazardous for other methods | For getting accurate temperatures, emissivity must be determined correctly |
| 3 | It can be used to find defects in shafts, pipes, and other metal or plastic parts | Most cameras have ±2% accuracy or worse and are not as accurate as contact methods |
| 4 | It can be used to see better in darker or completely dark environments | Thermography is only able to directly detect surface temperatures |
| 5 | It makes catching moving targets in realtime easy | – |
| 6 | It is a nondestructive test method | – |

(a) **Infrared Radiation**

Infrared radiation (Fig. 6.12) is electromagnetic radiation, such as radio waves, microwaves, ultraviolet rays, visible light, X-rays, and gamma rays. The wavelength of the infrared radiation "band" is 0.78–1000 μm (micrometers).

## *6.11.2 Tan-Delta Testing Kit*

Electrical properties of any insulating system change by ageing and continuous stresses. The insulation failure of any HV equipment is the principal contributor to the unexpected breakdowns. Insulating material in all the other materials in the equipment is most prone to stresses including thermal, electrical, mechanical, and environmental stress.

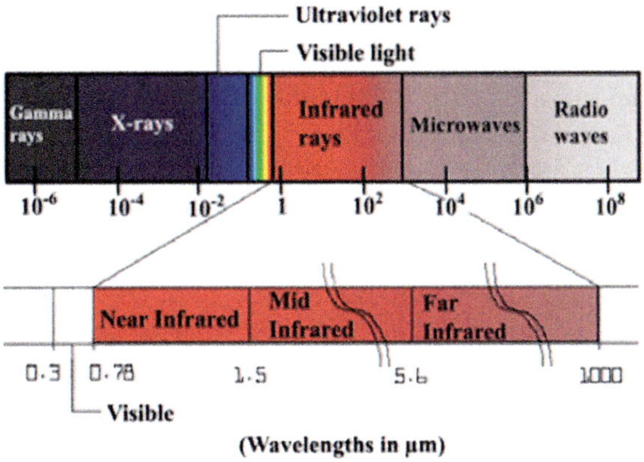


**Fig. 6.12** Spectrum of electromagnetic radiation

To try obviating the unexpected operational breakdown, periodic measuring of the electrical properties such as capacitance and tan-delta is necessary [9]. One of the most powerful nondestructive offline diagnostic tools to monitor the condition of solid insulation is the dissipation factor (tan-delta).

Capacitance and tan-delta values obtained on new insulation are treated as benchmark readings. Then periodic measurement and analysis of the readings with the benchmark readings can help to determine the rate of deterioration of the insulation's health.

Software Requirement

- Windows based software for measurement, data storage, transfer and analysis
- Expert diagnostic / expert guidelines software shall be provided for analysis of the data
- Shall measure Tan delta and power factor of all impedances (capacitive, resistive and inductive)
- Automatic time and date recording
- Digital display of all parameters Test voltage, Current, Tan delta, Power factor, Watts, Inductance, Capacitance, Resistance, Power loss, Frequency
- Real-time display of Power, Voltage, and Current.
- All dedicated software shall be supplied in original in CDs

### 6.11.3 *Cable Fault Detector Kit*

The main purpose of this kit (Fig. 6.13) is to find the fault in the length of underground cable from the base station in kilometers. The underground cable system is commonly used in low- and medium-voltage distribution lines. The advantages lie in its immunity from weather conditions such as pollution, rain, storm, snow, and ice. When a fault occurs, particularly for repairing underground cable, finding the fault location is a tedious process. The purpose is to find the exact location of the fault. The possibility of determining the gamut of faults with different fault characteristics rests on suitable measuring equipment and the operator's skills. The right combination enables reducing the critical fault outage time.

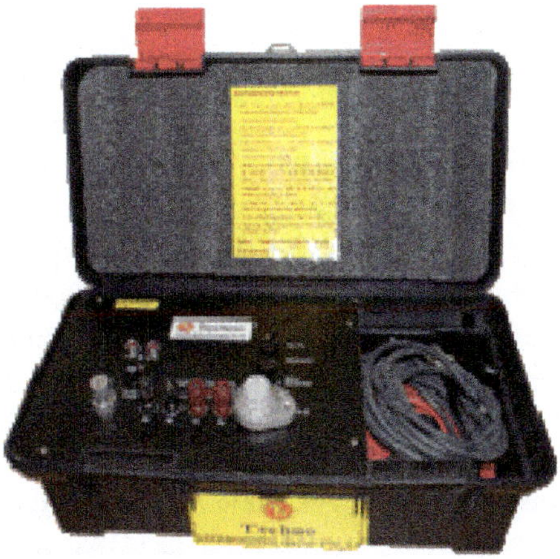

**Fig. 6.13** Cable fault detector kit

## 6.11.4 *Insulation Resistance (IR) Measurement*

One of the important tests is finding the insulation resistance of the electrical equipment for safety operations. Specifications of insulation resistance are displayed in connecting cables, motors, generators, protection devices, and the like, in order to limit the current flow outside the device or the conductors. The lifetime or the IR quality may change over a period of time due to electrical stress linked with under-/overvoltages, mechanical stress due to starting and stopping of electrical equipment, for the case of a motor/generator balancing problem, chemical stress, temperature variation, environmental contamination, and so on. The above causes reduce the resistivity of insulation and thus a leakage current increase in the circuit/equipment which may lead to serious safety problems. These tests help to deduct ageing and premature weakening of the insulating properties before reaching the incident described above.

An IR test applies a smaller amplitude of DC voltage (typically, 250 V DC, 500 V DC for LV equipment, and for HV equipment <600 V DC are used) compared to the amplitude of DC voltage used for the dielectric strength test and it measures the insulation resistance in kΩ, MΩ, GΩ, or TΩ when the phase and neutral are short-circuited together. These measured values indicate the resistance conditions; the higher the resistance values are, the better the condition of the insulation. Normally for ideal conditions the IR value should be infinite; practically it is not possible due to small leakage of the current.

**Testing Methods**

(a) **Short-Time or Instantaneous Measurement**

This is the simplest method, but its main disadvantage is the measured values are influenced by the temperature and humidity, possibly causing this measurement to be inaccurate. Therefore the measurement should be standardized at a specific temperature and the humidity level should be noted for comparison with previous measurements. It involves applying the DC test voltage for 30 or 60 s duration for noting the IR at the moment.

(b) **Polarization Index (PI)**

The successive IR values are calculated at specific intervals, 1 and 10 min durations, respectively. The ratio of the 10 min IR value over the 1 min IR value is called the polarization index; this value is mainly used to assess the quality of insulation. The advantage of this measurement method is that these values are not influenced by temperature as long as the equipment is not significantly subject to temperature variation during the test. The IEEE 43-2000 recommendation for rotating machines suggests that when the PI is less than 2, there is a problem in insulation; if the PI value lies between 2 and 4 the machine insulation is in fairly good condition and when the PI value is greater than 4, the machine has very good/ excellent IR value.

(c) **Dielectric Absorption Ratio (DAR)**

The DAR is defined as the ratio of the 60 s IR value over the 30 s IR value. When the DAR value is less than 1.25, there is a problem and insufficient insulation; if the DAR value is less than 1.6 the machine insulation is in fairly good condition and if the DAR value is greater than 1.6, the machine has very good/excellent IR value.

### *6.11.5 Megger Insulation Tester*

The Megger Company was the oldest IR test equipment manufacturer over 100 years ago (Fig. 6.14). This tester helps to find the IR value, which helps to verify the electrical insulation. The minimum values for the IR test vary; depending on the equipment type and the nominal voltage, it may vary according to international standards. Some standards will define the minimum IR test values for general electrical insulation. For example, for low-voltage insulation in the IEC world, IEC 60364 gives the minimum IR values, and recommended test voltage. Please see Table 6.3.

## 6.12 Gas Leakage Detector

The gas detection technique has a wide range of potential uses in electrical distribution and chemical industries. A gas leak detector (Fig. 6.15) has various sensors by which it detects and identifies potentially hazardous gas leaks by means of an audible alarm to alert people. Sensors used today include infrared point sensors (IPS), ultrasonic gas detectors (UGD), electrochemical gas detectors, and some

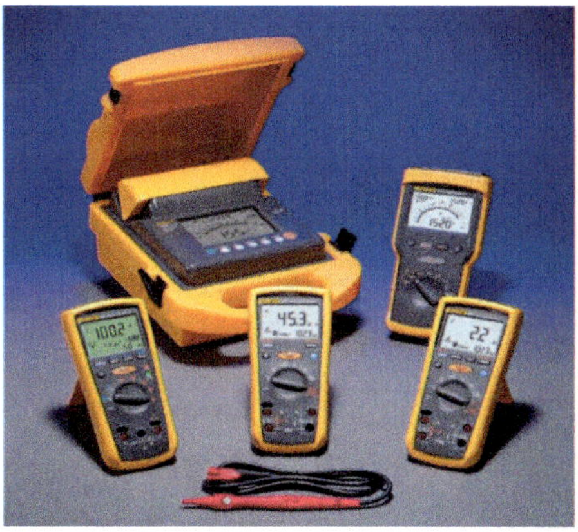

Fig. 6.14 Megger

**Table 6.3** Recommended test DC voltage for an AC circuit and IR value

| S. No. | Rated circuit voltage | Recommended test dc voltage (V) | Minimum value of recommended IR value in MΩ |
|---|---|---|---|
| 1. | Low-Voltage Circuit | 250 | $\geq 0.5$ |
| 2. | Up to 500 V | 500 | $\geq 1.0$ |
| 3. | Above 500 V | 1000 | $\geq 1.0$ |

other semiconductor sensors. These sensors are used for a range of applications in power plants, industrial plants, refineries, and wastewater treatment facilities. SF6 gas is 24,000 times more environmentally dangerous than $CO_2$.

## 6.13 Crimping Machine

Crimping tools are used to connect two pieces of metal or other adjustable material by cutting, stripping, bending, and deforming one or both of them to hold the other electrical and metal components. Crimping was developed as a high-quality, low-cost replacement for soldered termination [10]. It is recommended where soldering has been estimated as too expensive, complex, or time consuming to install. There is overall less processing, and the connection will be durable due to strain relief. It is essential to choose the right type of sleeve for the crimping tool. There are three common crimp sleeve types used in practice:

- **Round section sleeve**: It is made of brass and is used for making knots.
- **Oval section sleeve**: It is the most widely used type of sleeve, available in brass, copper alloys, and aluminum material.
- **Double barrel sleeve**: It is available in copper, brass, and aluminum material; it is very durable but most expensive.

**Fig. 6.15** SF6 gas leakage detector

The following types of crimping tools are used in the power sector (Fig. 6.16).

- **Hydraulic crimping tool**: It is hand operated and pumps hydraulic fluid into the device to compress the die. Crimping effort is significantly less as compared to other hand-driven manual crimpers.
- **Pneumatic crimping tool**: Air-filled crimp tools provide an easy and flexible crimping solution. These crimpers are fastened to an air supply, and perhaps an operational switch. Using these tools crimping can be done at a faster rate with high accuracy and efficiency, approximately 600 connections per hour.
- **Battery-powered crimping tool**: It has a motor controlled by an microcontroller unit (MCU), and powered by Li-ion batteries with a high-pressure hydraulic system and is extensively used on electrical construction sites.

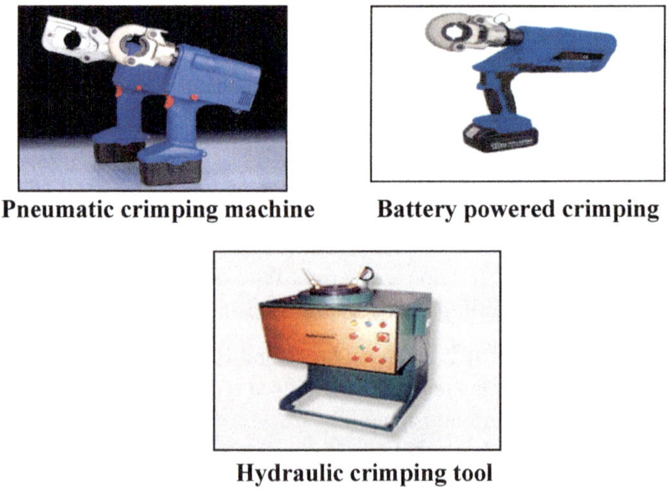


**Fig. 6.16** Crimping tools

## 6.14 Oil Filtration Machine

The oil filtration machine (Fig. 6.17) enables a process to remove oil contamination, sludge, dissolved moisture, and gases from the transformer oil. This process improves the insulation property; better insulation leads to longer life of the transformer and lessens the breakdown of the transformer. As a result there is a good return on investment of the transformer. Oil filtration machines are designed and manufactured per BIS, NEMA, IEEE, IEE, ASTM, and IEC standards.

### *6.14.1 Oil Purification Procedure*

The transformer is the most important property in an electrical power system; its life mainly depends on the oil quality. A good oil purification process delivers filtered oil with parameters as per standards. However, oil purification is based on the unprocessed oil standard. If unprocessed oil does not match the standards, then the oil purification process efforts are wasted.

(a) **Step 1**: The oil temperature is raised up to 65 °C. This assistance gives the oil latent heat which later aids in separating the moisture and gases from the oil in the degassing chamber. Also the oil viscosity drops to a certain level which helps for better filtration to some extent. The heating system is protected against overpressure buildup and excess temperature rise.

**Fig. 6.17** Oil filtration machine

(b) **Step 2**: Removal of sludge and dirt from the oil. There are two methods used for removal of sludge: filter candles and centrifuging action.

(i) **Filter candles**: Filter candles are the cartridge-type filtration in the oil purification machine and can be further classified into two categories:

- Using classical edge filter
- Using the depth-type filter

The edge filter cartridge or depth-type can be cleaned and reused at least three or four times, but cleaning and fitting involve considerable time and human effort. It is difficult to handle a large quantity of sludge. In order to remove the contamination in an edge or depth filter cartridge reverse-pressurized dry air or nitrogen flow used.

(ii) **Centrifugal action**: This is the second alternate method to remove the sludge. It is used for the separation of dirt from the oil and can be slow- speed-type centrifuges that do not additionally require an electric motor to spin the centrifugal cone at high speed. This method can remove the dirt at more than the 10-micron level; at less than 10 microns this method is not recommendable.

(c) **Step 3**: Dehydration and degasification of transformer oil are processes of dehumidification of the transformer oil and removal of gases is executed in the designing chamber. This dissolved water–oil separation or dissolved gas–oil separation is possible at reduced pressure, that is, vacuum, due to differences of the boiling point of water, gas, and transformer oil. In the process of separation of gases from the oil it becomes important that the aromatic hydrocarbons remain so that the original properties of the oil are retained. When the water level in the oil is above saturation level of the transformer oil, oil is observed in free water. Removal of free water can be done by power-driven centrifugal force or by the coalescing principle, where the latter is more effective and economical in practice.

### *6.14.2 Different Techniques for Removal of Solid Contaminants in Transformer Oil*

**Centrifugal Filter**

**Operating principles**: This filter (Fig. 6.18) accelerates the oil at very high speed. As a result centrifugal force throws the solid particles on the rotor wall and it is easy to remove the mass.

**Limitations**: This only removes the solid contamination up to 10 μm. The efficiency of this filter is low when it involves removal of solid contaminants in dissolved water, maintenance costs, high power demand, and capital.

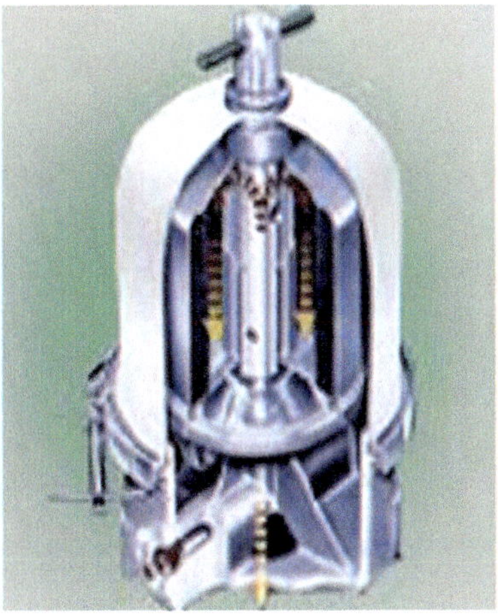

**Fig. 6.18** Centrifugal filter

**Fig. 6.19** Mechanical filter

**Mechanical Filter**

**Operating principles**: A filtration element for suspended contaminants made up of layers of media is used (Fig. 6.19).

**Limitations**: These filters can remove particles of 3 μm; however, they need periodic replacement or cleaning of their media.

**Magnetic Filter**

**Operating principles**: Geometrically arranged magnets produce nonuniform flux field zones that collect magnetic iron and steel particles (Fig. 6.20).

**Limitations**: They can only remove ferromagnetic particles larger than 1 μm and are not suitable for water removal in free or dissolved forms.

**Electrostatic Filter**

**Operating principles**: They use electrostatic principles to draw contaminants out of oil and trap them on the collector surface, including tars and varnishes (Fig. 6.21).

Fig. 6.20 Magnetic filter

Fig. 6.21 Electrostatic filter

Advanced electrostatic filters can easily remove submicron solid particles and some soft contaminants.

**Limitations**: They have slow operation and loss in effectiveness with high moisture content and HV operating is needed.

### *6.14.3 Different Techniques for Water Removal*

(i) **Centrifugal Separator**

The working principle is the same as the centrifugal filter and can remove free and some of the emulsified water but is inefficient with dissolved water.

(ii) **Coalescer Separator**

The coalescer separates water droplets from the fluid stream by arresting them on a filter's surface by fusing droplets together to a size that allows them to fall into the vessel where they can be extracted. This technique does not allow separating dissolved water and a fine upstream filtration of any particulate contamination.

(iii) **Absorbent Filter**

It is suitable to remove free and emulsified water by superabsorbent polymers impregnated in the filter matrix.

**Vacuum Dehydration**

In a partial vacuum dry hydraulic oils and lubricants are exposed where the concentration gradient is utilized between the fluid and the evacuated air to evaporate the water from the fluid.

There are main techniques: flash distillation vacuum dehydration and mass transfer vacuum dehydration. The main drawback of the first is that the high temperature and vacuum employed can lead to loss of lower boiling base stock fractions and volatile additives and can result in thermo-oxidative oil degradation. The mass transfer dehydration removes all free water. However, up to 30% of dissolved water remains in the oil.

## References

1. Steward WE, Stubbs TA (1995) Modern wiring practice-design and installation. Newness, U.K.
2. Tripathy SC (1991) Electric energy utilizations and conservations. Tata McGraw Hill, New Delhi, pp 73–105
3. IEEE Standard 446-1987, IEEE Recommended Practices for Emergency and Standby Power Systems for Industrial and Commercial Applications (IEEE Orange Book) (ANSI)
4. IEEE Standard 525-1987, IEEE Guide for the design and installation of cable systems in substations (ANSI)

5. IEEE Std 242-1986 (Reaff 1991), IEEE Recommended Practice for Protection and Coordination of Industrial and Commercial Power systems (IEEE Buff Book) (ANSI), Chapter-4
6. ANSI/NFPA 70-1993, National Electrical Code
7. ANSI/NFPA 110-1993, Emergency and Standby Power Systems
8. ANSI/UL 508-1988, Safety Standards for Industrial Control Equipment
9. Beeman DL, (ed) (1995) Industrial power systems handbook. Tata McGraw Hill, New York, Chapter-2
10. IEEE Std 979-1984, IEEE Guide for Substation Fire Protection (ANSI)

# Chapter 7
# Best Practices in Distribution Engineering and Automation

## 7.1 High-Voltage Distribution System

For losses in the distribution system and for quality improvement of the supply high-voltage distribution system (HVDS) system it is recommended to erect a 5 kV, 10 kV, or 16 kV small-size single-phase transformer with 11 kV lines nearer to the load and release the supply at least to the LT line. Aerial bunched cables (ABC) cover short LT lengths.

### *7.1.1 Technical Advantage of HVDS Lines*

Figure 7.1 shows a HVDS rural electrification.

- Improves voltage profile.
- Reduces losses.
- Avoids frequent fuse blowouts.
- Fewer percentages of transformer failures.
- Uses smaller conductor size.
- Quick replacement.
- Due to distribution transformer failure affects few consumers.
- Improves power factor (normally >0.95) initiating reactive power control.
- Normally LT lines are insulated and length is small thus it avoids unauthorized hooking of loads.
- Each DTR contains few pumps connected to it and the consumers conclude ownership and liability.
- Enhances power quality and experiences consumer satisfaction.



J. Raja et al., *Practices in Power System Management in India*, Power Systems, https://doi.org/10.1007/978-981-10-2972-1_7

**Fig. 7.1** HVDS used in rural electrification

### 7.1.2 *Critical Issues Frequently Faced on HVDS*

- Is HVDS meant to electrify future areas or for converting existing LVDS?
- Do HVDS and LVDS exist together in the same region?
- What is the connection between load density and kind of distribution system?
- Economics and payback period.
- Guidelines on acquiring single-phase transformers.

If three-phase pump sets are used, HVDS can be applied by converting existing LT lines for 11 kV and three smaller size single-phase transformers can be used to provide three-phase LT supply capacity. Using one or two bushings, Single phase transformers 6.3 kV/233 V can be constructed. HVs are used to connect three single-phase transformers in star or delta [1, 2].

**Linkage with Current Density**

- High-load density area where the load capacity exceeds 20 MW/km$^2$ such as metro areas due to increase in multistoried complexes; in practice LT alone is not able to carry much load.
- In smaller urban areas, three single-phase transformers are employed in place of a three-phase transformer. Loads on a limb can be distributed to the other two limbs in the event of failure of one limb. Also a failed transformer can be easily transported and erected.
- The entire unit should be replaced in the case of three-phase large capacity transformers, and downtime is high.

- In rural areas, one or three single-phase transformers can be employed without any doubt.
- Loads in rural areas and even village habitat portions are too low and generally single phase.
- In villages, the load densities are confined to be very low and thus are electrified afresh.

**Economics**

The cost of a three-phase transformer is 1.3 times the cost of three single-phase transformers and the capacity of a three-phase transformer is equal to the capacity of three single-phase transformers [3]. Also no-load to full-load losses are almost the same. However, as compared to LT, current from the same capacity is too low at 11 kV. Current is 5 amps at 11 kV and 140 amps at LT for a 100 KVA line. Reduction line losses are in the ratio 52:1402 (25:19,600). Thus, it is very inexpensive.

### 7.1.3 *Three Single-Phase Versus Single Three-Phase Transformer in Urban Area*

The national energy policy government of India targets "A lamp for each house" under the scheme of *Kutir Jyothi* for all rural areas. To implement this scheme high capacity transformers are normally not needed. In light-load areas even the use of single-phase transformers with capacities of 5 MVA, 10 MVA, 15 MVA, or 16 MVA are more advantageous as follows.

(i) Single-phase loads can be connected on individual transformers suitably dividing them. For an extreme case, 4 kW or above capacity loads can be connected with the smaller capacity of a three-phase transformer. If loads are very low, they cannot contribute and create high imbalance.
(ii) Loads can be distributed to the other two limbs in case of failure of one limb.
(iii) Transport and replacement are very easy and can be done much faster. In fact a rolling stock of 4% can be maintained at each distribution section office for faster replacement.

### 7.1.4 *Restrictions Regarding LVDS and HVDS*

- Domestic customers constitute 99.99% of existing customers. Their operating voltage is 415/240 volts. Hence, the key to customer services is nothing but operational performance of the LV network.
- In the Indian power system T and D are generally more than 20%. In particular, the LV system is responsible for high losses; per the study, LV line losses are six times the target limit and three times the maximum tolerable limit. Switchover to HVDS alone can bring losses to international norms.

- The ratio of power loss for transmitting an equal load in LVDS (415 V) and HVDS (11,000 V) lines is 13:1.
- Voltage drop between distribution substations and the customer region should not be allowed more than 10%. Hence, in accordance with IEC Rule 56 on voltage drop, it is complex and very costly in LVDS whereas in HVDS, it becomes very easy.
- While investigating typical LV feeders in LVDS, it was indicated that 75% of LV feeders have a volt drop above 5% which causes high losses. But in HVDS losses on the LV line are negligible.
- Existing conductors in LVDS are loaded above economic loading limits because the current for distribution of the same power in LVDS is high. It can be avoided with a switchover to HVDS.
- The number of feeders to be monitored in LVDS and HVDS are in the ratio of 60:1. Therefore the monitoring of feeders in LVDS is much more burdensome compared to HVDS.
- In LV lines, unofficial tapping is easy and uncontrolled in LVDS whereas it is very difficult in HVDS.
- The cables are locally available and are manufactured with tough insulation by ABC cables.

### *7.1.5 Three-Phase HVDS*

Under a restructuring distribution scheme, the existing low-voltage lines are converted to single-phase two-wire HV lines. Three-phase distribution transformers with small capacity single-phase transformers are also replaced [4].

**Advantages of HVDS System**

- The feeling of ownership by listed customers would boost them to take responsibility. As a result they would not allow others to meddle with the LT network.
- Unofficial loads are anticipated by the consumers themselves inasmuch as loading beyond capacity may cause the distribution transformer to fail.
- Underloading and no tapping of LT lines will minimize the failure.
- Equipment failure only affects two or three customers instead of 25–30 customers in the original system.
- No voltage drop leads to high quality of supply.
- Good voltage and less fluctuation lead to less burnout of motors.
- Saves power purchase cost because of reduction in losses.
- Less loss leads to supplying more loads without any further investment on infrastructure.
- The breaker trips at the substation as the line is at 11 kV potential; hence it leads to reduction of accidents due to touching of snapped conductors.

## 7.2 Aerial Bunch Conductor (ABC)—Based Distribution System

For an overhead power distribution network, aerial bunched cables are a new concept compared to the conventional bare conductor. They are used to provide safer and more reliable operation of the system [5]. The economics of the overall system as power losses are reduced and the costs of maintenance and operation are also reduced. This kind of system is very beneficial for installation at difficult locations such as hilly, forest, and coastal areas, and it is ideal for rural distribution. For power distribution in urban congested areas with small lanes, ABC can be considered the best choice. ABC is a better choice in developing an urban complex owing to its flexibility for diversion because demands keep changing in an urban development plan.

### *7.2.1 Constructing ABC*

Power conductors are made up of aluminum (neutral conductor and street lighting conductor if and when necessary), insulated with XLPE/HDPE. These power conductors are twisted around high tensile standard and galvanized steel (aluminum alloy may be used) in either insulated or bare messenger wire in order to make the aerial bunched cable. This assembly is hung directly to the distribution pole/towers with the help of standard hardware available on the market but it should be taken into consideration that the restored messenger wire is completely insulated in case of HT ABC from earthing at any point of distribution. The insulation of XLPE (cross-linked polyethylene) is black in color. Due to exposure to direct sunlight and ultraviolet radiation, insulation needs to be stabilized against deterioration; thus insulation is black in color. XLPE is made thermoset by special formulation from a base polymer of thermoplastic low-density polyethylene. It is a low-density polyethylene that is cross-linkable. XLPE merges the best electrical properties of LPDE along with superior thermomechanical properties.

### *7.2.2 Components*

- Aluminum conductors abide by IS:8130 (Class-II).
- Standard high-tensile galvanized steel messenger wire obeys IS:398 (Part—2). But alloy messenger wire obeys IS:398 (Part—4).
- XLPE insulation specifies IS:7098 (Parts—I and II) and HDPE insulation of power conductors specifies IS:6474, respectively.
- With the introduction of messenger wire, the tension from the current-carrying conductor is totally removed and the operating temperature of the conductor is lowered from 90 to 75 °C of the bare conductor of the same size.

### 7.2.3 Applications

**Uses for ABC**

- As a substitute for uncovered and unprotected transmission lines in places such as in rural areas, in woods, and in other localities and narrow streets where there is limited space.
- As a substitute for bare transmission lines where reliability is most important.
- As a substitute for bare lines where a high degree of stability of supply voltage is required.
- Where the cost of a joint of overhead lines and underground cables becomes very high such as in hilly terrains.
- As augmentation of the present system without actually increasing voltage.
- In many temporary supplies.

### 7.2.4 Advantages

- Less fault rate because of superior protection against both line and ground fault caused by high winds, falling trees, and birds mainly in hilly areas and forests as seen in rural distribution networks.
- High resistance between conductors and between conductor to earth in all seasons and even in polluted atmospheres.
- Low losses and negligible leakage currents.
- Power and telephone cables may be wound in the same set of poles or any other support such as walls.
- Better flexibility to run parallel with existing overhead bare conductor system without any intrusion.
- Impedance of the line is low because of high capacitance and low inductance of lines.
- Lower voltage drop, higher current capacities.
- Better voltage regulation, much safer than bare conductors, easy maintenance, and cost of lines is reduced.

## 7.3 Consumer Indexing

In a large area in a highly populated country like India, the distribution and utility network is complex [6, 7]. It is also vast and scattered over a large area that is divided into manageable administrative boundaries. With growth in the overall area, it becomes more complex and grows faster over time. Because there is no managed and updated database, service connections and consumer data are not documented properly and become very difficult to locate. Surveys have shown there is no need to find illicit connections.

There is a requirement of proper documentation of distributed networks over a geographic base map for the effective management of the vast distributed network with huge assets. Mapping for the number of assets and consumers is done through surveys and indexing. In accordance with distribution network criteria and standards, there is also a requirement in order to verify the electrical documentation. In its true sense, this can be used to comprehend a simple drawing over a geographical information system (GIS) platform and electrical network documentation. To meet the need of GIS documentation and consumer indexing in electricity transmission and distribution network, integrated software with features and functionalities for power utilities is necessary. These transmission and distribution networks contain substations, switching stations, HT/LT feeder/branch, power and distribution transformers, towers/poles, junction box/pillar box, and many other electrical elements that jointly constitute the power system.

Geographical co-ordinates are used to plot an object over a geographical base map ensuring its location. Either manual plotting or capturing the GPS co-ordinates is used to achieve this. The requirements include documentation of consumers, different voltage level connections, and meters along with their association within the network. Core modules of this software include GIS and single-line diagram (SLD) documentation, consumer mapping, and indexing. There is also an optional feature offered by the software comprising mapping of consumer plots in the form of parcels. The product allows deployment in many areas as it has the potential to integrate with a third party with the help of an exposed B2B interface. As the base product becomes easy to customize, the software would be able to meet the needs of specific customers.

The documentation that would meet the standards for electricity transmission and distribution networks is ensured by the software which gives on-draw validation. Presentation supports a layerwise view. The indexing format can be adapted to specific definitions and can also be redefined and regenerated in order to make the indexing format fully flexible and user customizable.

## 7.4 Up-to-Date Asset Register

Energy companies own and operate the infrastructure and assets that are in a constant state of flux. Network assets can be continuously added, moved, or replaced and this can be done either directly by the utility or by subcontractors or other agencies. This work is generally planned but it can also be done in order to maintain system integrity in response to an emergency [8, 9].

It is essential for each utility to record up-to-date assets. For the purpose of tracking assets use of GIS technology will be beneficial.

GIS-based asset mapping is a methodology in which a satellite image is digitized for the area of interest. The same map is loaded over portable GIS survey equipment (such as a palmtop). A surveyor with that base map visits each asset and records details of the asset along with its GIS coordinates. After the survey it is uploaded to the database and a list of assets/asset register is prepared. It is essential to update the asset register on a regular basis at least once a month.

There are various benefits for maintaining such records:

- List of total assets of the utility, ageing, costs, and so on can be calculated. Based on this, a future maintenance/augmentation schedule can be planned.
- In the case of some faults it is easier to locate the actual fault position and impact. However, attending faults will be easier.
- Based on this record the network flow can be traced. Energy/technical loss can be calculated. Distribution automation can be implemented based on this information.

## 7.5 Distribution Automation

Nonavailability of distribution network topological information and current status of the equipment such as distribution transformers, capacitors, switches, and feeders hampers the O&M of a distribution system [10, 11]. Operational as well as future network expansion planning cannot be done concurrently. Delayed fault detection, isolation, and service restoration are common features of a weak distribution network. All these lead to increased system losses, poor quality and reliability of the power supply, in addition to the increased peak demand and poor return of revenue. The situation can be corrected by intervention of IT initiatives, GIS mapping, and adhering to DA (distribution automation).

Automation refers to doing a particular task automatically and at a faster rate in a sequential manner. This can be achieved by use of a microprocessor in co-ordination with a communication network and some relevant software programming.

Application of automation at the distribution power system level can be defined as automatically monitoring, protecting, and controlling switching operations through intelligent electronic devices (IED) to restore power service during a fault by sequential events and maintaining better operating conditions back to normal operations. Recently, due to advancements in communications technology, a distribution automation system (DAS) is not just the remote control and operation of substation and feeder equipment but results in a highly reliable, self-healing power system that responds rapidly to realtime events with appropriate actions. Hence, automation does not just replace manual procedures; it permits the power system to operate in the best optimal way, based on accurate information provided in a timely manner to the decision-making applications and devices.

**Summary of DAS Primary Aims**

- To optimize the operational and energy efficiency of distribution networks, and improve the quality and reliability of the power supply.
- To form an integral part of the evolving smart grid by deploying intelligent distribution network elements and devices that can communicate with each

other, and with other downstream and upstream elements, thus furthering the smart grid objectives of flexible, responsive, and efficient management of the power supply in line with demand.

Distribution utilities in their effort to automate existing substations and other components of a distribution network should focus on two aspects—economical and technical—that influence the optimum control of the power system management business.

**Economic**

- Cost reduction in operation: Faster fault location, clearance, and shorter supply interruption through better co-ordinated network control can lead to personnel and other related cost reductions.
- Cost reduction in maintenance: Reduction in maintenance cost of primary network components by changing from presently followed preventive, (time-based) maintenance to predictive and RCM (reliability-centered maintenance).
- Installation cost reduction: By adhering to retrofitting of existing equipment to make it automation compatible, savings in cost can be achieved. Space saving and control room building cost saving can be achieved for substations.

**Technical**

- Online information of system parameters and consumer loads would be useful for online monitoring of system stability.
- Documentation of such data would be useful in future and timely planning and engineering exercise.
- New functionalities can be added to the existing equipment.
- Setting and co-ordination of relays can be done online.

Thus in addition to the tangible benefits of improving reliability and efficiency within network operations, DA upgrades have another important attribute: they have the potential to deliver a strong return on investment without requiring intensive consumer engagement or behavior change. DA technologies begin to grasp the demand response capability of conservation voltage control (CVR) and the energy management possibilities of dynamic load distribution, once it is sustained with relatively simple remote control of field-based switches having the main goal of increasing reliability.

Apart from a supervisory control and data acquisition (SCADA) and distribution management system (DMS), implementation of a geographical information system (GIS), automated meter reading (AMR) system, outage management system, communication system, billing and business process automation, and a SAP-based enterprise resource planning (ERP) system are possible.

In order to meet the requirements to transform conventional static grids into modern and dynamic smart grids, distribution networks are under very high pressure. This trend is influenced by the increasing occurrence of decentralized generation (DER) particularly and apart from that, the need to improve the quality and reliability in MV and LV networks. Along with the requirements on operation and maintenance ever-growing cost pressure is an add-on. New requirements have arisen for automation, monitoring control, and protection of distribution substations and ring main units (RMUs) because of the shift in paradigm. Consistent and flexible system solutions, which are scalable for different applications, are used to support these requirements. In order to ensure seamless and interoperable systems in communication, there should be consistent use of standards. The segments' generation, transmission, and subtransmission perform at a high level in today's electrical grids and these also have substation automation systems. But the distribution segment and distribution substations (feeder heads) along with the ring main units, are positioned along the feeder and thus far only have a basic level of communication and automation. Most of the existing RMUs are not equipped with any intelligence for communication and automation. Furthermore, there may be a variation in the levels of automation in different countries due to previous strategies of the utilities. Another critical area in many distribution grids may be a lack of space in RMU housing for the necessary peripherals of the feeder automation system. Finally, there exist a large variety and differences in the age of primary equipment, which is not prepared for advanced automation and communication; for example, many RMUs are not prepared for remote control, or a fuse is used for protection purposes instead of a circuit breaker. Therefore, the smart grid policy of the utility companies requires an individual migration and modernization strategy for a future-oriented distribution automation and protection solution.

First of all a concept has to be designed which states that for the distribution substations and RMU, there is a required level of automation and functionality. Because of the presence of distinct primary equipment or communications availability, there could have been a difference in the RMU in one distribution grid or in the same feeder. At the same time, a certain level of automation and smart grid functionality could be realized with or without limited access to communication. There can also be a mix of functions in one feeder automation system.

The roadmap for grid upgrades towards a smart grid can be provided by the following levels of distribution automation.

**Local Automation (Without Transmission)**

- Sectionalizer (by using switching sequences, automated restoration of fault)
- Voltage regulator (regulation for long feeders, automation of voltage regulation)
- Recloser controller (for overhead lines, automatic closing of breaker after use).

**Monitoring Only (One-Way Transmission to Distribution Substation or Control Center)**

- Messaging box (e.g., for fast fault location, short circuit indicators with one-way communication to distribution substation or control center).

**Control, Monitoring, and Automation (Two-Way Transmission to Distribution Substation or Control Center)**

- DA-RTU (distribution automation RTU) with powerful communication and automation features and applicable for smart grid functions.

## 7.6 DMS (Distribution Management System)

Automation involves "A set of technologies that enable an electric utility to remotely monitor, coordinate and operate distribution components in a real-time mode from remote locations." Please see Table 7.1.

The selection of distribution automation plays a key role in maintaining the power network with improved operation and maintenance, high reliability, improved efficiency, and power quality [12].

The substation automation system (SAS) is the latest technology trend that provides control of all the transmission and distribution substation equipment from remote control centers (RCC), as well as from the local control center, along with monitoring. Local station control is enabled by SAS using a PC by means of a human–machine interface (HMI) and control software packages containing an extensive range of functions by SCADA. For the control of bay and inter-IED communication infrastructure, a communication gateway and intelligent electronic devices (IEDs) should be included. The SAS adheres to IEC 61850 standards.

DMS performs many functions such as managing the operation of the electrical distribution network and the field crews assigned to operate, maintain, and repair the network. Improvement of the utilization of the distribution network and increase in productivity of the workforce are some of the functions of a DMS network manager. Thus the flow of data and information among operations, engineering, management, and customers is enhanced.

**Table 7.1** Function of automation

| Substation automation functions | Feeder automation functions | Consumer-level automation functions |
|---|---|---|
| Data acquisition and supervisory control of<br>• Circuit breakers<br>• Load tap changers<br>• Capacitor banks<br>• Transformers<br>• Fault location<br>• Fault isolation<br>• Service restoration<br>• Substation reactive power control | Data acquisition and supervisory control of<br>• Line reclosers<br>• Voltage regulators<br>• Capacitor banks<br>• Sectionalizers<br>• Line switches<br>• Fault indicators<br>• Fault location<br>• Fault isolation<br>• Service restoration<br>• Feeder reconfiguration<br>• Feeder reactive power control | • Automatic meter reading<br>• Remote reprogramming of time-of-use (TOU)<br>• Meters remote service<br>• Connect/disconnect<br>• Automated customer<br>• Claims analysis |

A connectivity model of the as-built electrical network is maintained completely by the system.

The system is able to retain an accurate representation of the as-operated phase of the electrical network by processing inputs from EMS/SCADA (see Table 7.2).

The most efficient manner to analyze the status of a network in terms of loading (current and voltage), outages, fault location, and dispatch crews, other information inputs such as substation loads, customer trouble calls, relay fault data, and crew location can be used [13]. In this way, the DMS network manager can provide a more accurate picture of the electrical distribution network performance as compared to a standalone SCADA system. It can also allow for more reliable responses to system disturbances at improved speed.

The following are the functional requirements to be built into a distribution management system.

- Common user interface (SCADA, DMS, GIS)
- Historical information system
- Network topology processor (automated mapping and facilities management)
- Outage management
- Switching order management
- Fault location, isolation, and service restoration
- Load balancing
- Interfacing to trouble call system (complaint system)
- Crew management
- Load management (auto and on-demand scheduled load shedding)
- Optimal feeder reconfiguration
- Distribution power flow
- Load survey and energy accounting.

For control center operators, service center and call center personnel and management, full graphics and Web-based interfaces are being offered by the DMS network manager. To support various work procedures such as the requirement for switch orders, trouble calls, crew dispatch, outage management, and to allow access to up-to-the-minute operational status reports, these are expanded throughout the utility.

A typical distribution network consists of following components.

**Table 7.2** Assessment results

| Global level | Indian utilities |
|---|---|
| IT used to enable operations at transaction level to provide benefits:<br>• In-built process control<br>• Workflow-enabled transactions<br>• Single point of data capture<br>• Support for timely strategic decision making | Core operations are manual and faces issues such as<br>• Ad hoc decision making<br>• Poor data quality<br>• Long decision-making cycles<br>• Underutilization of IT investments |

**Transformers**

- Circuit breakers
- Feeders
- Sectionalizing switches
- Capacitor banks
- Voltage regulators
- Small generation sets
- LT consumers
- HT consumers.

Automation could be done at three levels.

- Substation level automation
- Feeder level automation
- Customer level automation.

Automation can address the following types of problems and other jobs.
Fault location, isolation, and service restoration.

- Maintaining good voltage profile
- Load balancing
- Load control
- Metering
- Maintaining maps
- Fuse-off call operations
- Energy accounting
- Outage management
- Customer information system management.

Distribution automation requires substantial IT interventions and communication devices/systems to be deployed by distribution utilities.

This wide gap between Indian and global best utilities needs to be bridged to improve commercial and operational performance of the utilities including quality of service to consumers. With the finances made available under the R-APDRP project by the Indian government, Indian utilities have launched massive programs of installing the required IT hardware/software. Please see Table 7.3.

**Communication Technologies**

UMTS: Universal mobile telecommunication system (3G system); WiMAX (worldwide interoperability for microwave access professional mobile radio (or private mobile radio, PMR); point to multipoint communication using VHF/UHF frequency show the way forward towards smart grid [14].

Because the communication and information infrastructure is expected to be the backbone of a smart grid, electric utilities all over the world are aware of the opportunity and are already investing in them.

Collaborative smart grid demonstration projects are used to unveil the difference in standard and open communications. These are related to the integration of

**Table 7.3** Wireless communication technologies available to november 2011

| Technology | Standards | Operator owner | Freq. band | Data rate | Application |
|---|---|---|---|---|---|
| VHF/UHF radio | Proprietary PMR | Utility | 150/ 400 MHz | Narrowband | Voice, DA, and SCADA |
| 2.4 GHz wireless | WLAN ZigBee | Customer utility | 2.4 GHz | Broadband | Short-range AMR, home automation |
| Point to multipoint | Proprietary, WiMAX | Utility or third party | 5–60 GHz | Broadband | High-speed data, DA, SCADA |
| Public cellular data services | GSM/GPRS, UMTS, CDMA | Third party | 900/ 1800 MHz (EU) 800– 1900 MHz (EU) | Narrow band, broadband | Voice, data, DA, AMR |
| Satellite communication | Proprietary | Third party | 6, 12 GHz | Narrow band | AMR |

distributed resources and would get the smart grid on the "Slope of Enlightenment" shortly. The success of technical and financial investments is the area of interest of investors and regulators.

Customers are interested in understanding whether the profit that can finally be produced will justify the cost.

## 7.7 Mobile Substation

A mobile substation is used during war, natural catastrophe, or an equipment breakdown in order to bring electricity temporarily to regions that have lost it.

It constitutes the components such as the trailer, switchgear, breakers, emergency or station power supply, a compact high-power-density transformer, and raised cooling capability (Fig. 7.2). It reaches the blackout location within 24 h to provide temporary electrical power. Its capability to move from one area to another quickly is its greatest advantage [15].

Apart from mobility, it also offers service providers time to repair damaged parts and get normal service back. It is also useful when maintenance is needed on substations so that consumer's power does not shut off during work. They are made in compact size so that units become easier to use in smaller regions. Please see Fig. 7.3.

Utilities also use mobile transformers so that out-of-service transformers are replaced temporarily, either for maintenance or due to forced outage. These mobile transformers are available in medium power range (10–100 MVA) with HV rated at 245 kV.

**Fig. 7.2** Mobile substation components

**Fig. 7.3** Mobile substation in transit

### Some Potential Purposes of Mobile Substation

- **Planned maintenance:** A mobile substation is used on a day-to-day basis within the utility to provide alternate capacity during planned maintenance of substations. It is desirable to have mobile substation systems available for emergency duty during peak loading or extreme weather conditions; utilities schedule their planned maintenance around the time when mobile substation systems are less likely to be needed for emergency use.
- **Temporary substation capacity increase:** Mobile substations may be called upon when an area may be faced with a temporary load increase that is not expected to last more than several months. Examples are construction projects or major plant modifications that require high electrical loads that will drop following completion. Special events can boost the capacity needs for a short time period.

- **Forced outage repair:** Unplanned repairs can be called for due to existing equipment failure, weather phenomena, or intentional disruptions. Equipment failure is the most common cause for deployment [16].
- **Weather and other natural outages:** Disasters are the main cause of electrical outages, although most often these have a larger impact on the power lines leading to and from the substations than on the substations and transformers themselves. Some natural disasters can harm substation operations and create a need for mobile substation systems.
- **Sabotage and attacks:** Intentional disruptions such as sabotage could severely harm the electrical grid, and mostly substations are vulnerable to attack. These systems may be useful in returning the facility to normal operations more quickly. This may be especially true if the attack strikes several substations, perhaps in order to bring down portions of a large urban area.

## References

1. Computer Aided Distribution Management Utility Practices, Publication No. 227, CBIP, Malcha Marg, Channakyapuri, New Delhi
2. Wiebe M (1999) A guide to utility automation. Pannwell, USA
3. Stevenson A (1993, June) Which communication technology is best for automation? Transmission and distribution, pp 44–53
4. Peri R, Krishna Rao MV (1994) Load control for India. Transmission and Distribution International, Fourth Quarter, pp 26–30
5. Sehgal BK (1989) Computer aided electric power control. Power & Process Tech, Incorporated Publication Division, New Delhi
6. Asian Electricity (1999, Mar) pp 22–24
7. Asian Power (2000, Sep), pp 65–66
8. Berney GC Intelligent instrumentation. Prentice-Hall of India, New Delhi, pp 10, 118, 125
9. Pabla AS Future technologies for rural electrification in India. In: Proceeding of international R&D conference, CBIP, New Delhi 9–12, pp 339–351
10. Khare PN (1999) Load management by judicious load shedding. Electrical India, 28th Feb, pp 12–15
11. Strauss C (2003) Practical electrical network automation & communication system. Newnes, Amsterdam, pp 49–50, 69
12. Craig D (2010) Communication options for distribution automation and automatic metering. In: Proceedings of IEEE PES transmission & distribution conference, New Orleans, U.S.A., April 19–22
13. IEEE Guide for Power Line carrier Applications, IEEE Standards 643-2005
14. Marihart DJ (2001, Apr) Communication technology guidelines for EMS/SCADA systems. IEEE Trans Power 16(2):181–188
15. Gelblen L (1996, Oct) Distribution automation increases reliability. Transmission and Distribution World, pp 38–41
16. Bush R (1996, Jan) T&D spending patterns. Transmission and distribution, p 61

# Chapter 8
# Best Practices in Operation and Maintenance of Energy Meters

## 8.1 Metering Provisions

The energy meter is a device that records the energy consumed over a specific period of time. It is an integral part of revenue realization in the distribution business. All domestic and industrial consumers need at least one energy meter to register energy consumption. Based on the meter reading, the supplier calculates the bill to the consumer [1, 2]. The generation company sells the electricity to the distribution company which has the responsibility to sell this energy to the consumer and consumer needs to pay the bill from the distributor.

Accurate energy accounting remains elusive in the Indian power distribution system, as utilities continue to manage with unmetered consumer segments, incomplete transformer metering, and lack of baseline data. This leads to significant limitations in efficient determination and loss reduction.

## 8.2 Indian Electricity Act 2003

The Indian Electricity Act 2003 notified on June 10th, 2003 the objective of competition, protection of consumer's interest, and power for all consumers located in various regions. It creates a liberal framework for power development and focuses on revenue recovery and protection. Act 2003 creates an environment to run the power distribution business as profit centers as well as encourage private investment.

### *8.2.1 Electricity Act 2003 Section 55*

In Section 55, the Act 2003 amendment explains after the two years of expiration from the date of appointment, except through installation of the correct meter, the



J. Raja et al., *Practices in Power System Management in India*,
Power Systems, https://doi.org/10.1007/978-981-10-2972-1_8

licensee is not entitled to supply electricity in accordance with regulations. For proper accounting and audit in generation, transmission, and distribution or electricity trading, the authority may direct the installation of a meter by a GENCO or licensee at such stages of generation, transmission and distribution, or electricity trading, and at such locations it may be deemed necessary.

### 8.2.2 *Central Electricity Authority Notification Dated March 17th, 2006*

The Central Electricity Authority (CEA) notification provided for the classification of meter types, specification and standards, proprietorship, location, precision class, installation, testing, operation and maintenance (O&M), access, meter safety seal, reading and recording, failure or inconsistencies in the meter, tamperless features, quality assurance, calibration and periodic testing, and adoption of new technologies in respect to the following meters for correct accounting, billing, and audit of electricity.

- Interface meter
- Consumer meter
- Energy accounting and audit meter

All the above-specified meters should be in accord with the Bureau of Indian Standards (BIS) [3]. If equipment or material do not have any BIS Standards, the corresponding International Electro-Technical Commission (IEC) Standards, British Standards (BS), or any other equivalent standard should be practiced, provided that whenever an international standard is followed, necessary customization or alterations should be made for the parameters such as system frequency, voltage, ambient temperature, humidity, and other conditions prevailing in India before actual adoption of the said international standard; the standards on "Installation and Operation of Meters" as specified in the schedule annexed to these regulations and as amended from time to time should be followed.

## 8.3 Variety of Meters

Figure 8.1 depicts the different kinds of meters.

## 8.4 Location of Meters

Meter locations are shown in Table 8.1.

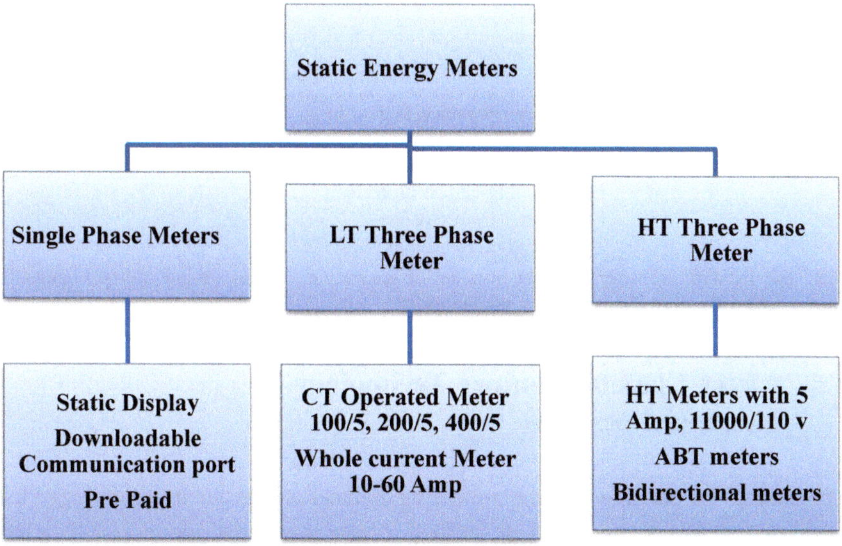


**Fig. 8.1** Meter varieties

**Table 8.1** Location of interface, consumer, energy accounting, and audit meters

| S. No. | Stages | Main meter | Check meter | Standby meter |
|---|---|---|---|---|
| 1. | Generating station | All outgoing feeders | All outgoing feeders | • HV side of generator transformer<br>• HV side of all station auxiliary transformers |
| 2. | Transmission and distribution system | • For same licensee, one end of the line between the substations<br>• For two licensees, both ends of the line between substations; for respective licensees, meters at both ends considered as main meters | – | • No separate standby meter<br>• In the case of two different licensees, meter installed at other end of the line works as standby meter |
| 3. | Interconnecting transformer (ICT) | HV side of ICT | – | LV side of ICT |
| 4. | Directly connecting consumers to the inter/intra-state transmission system, covered under ABT and permitted open access by the appropriate commission | As directed by appropriate commission | | |

(continued)

**Table 8.1** (continued)

| S. No. | Stages | Main meter | Check meter | Standby meter |
|---|---|---|---|---|
| 5. | Consumer energy meters (as per supply code) | Meter installed at a height enabling convenient meter reading and to protect meter from adverse weather conditions | – | – |

*Note* The location of meters already installed at generating stations should not be changed or altered without proper approval from the authority

## 8.5 Meter Classification per Technology, Use, and Locations

Meters can be classified in the following manner as per their technology, use, location, and so on.

### *8.5.1 Interface Meters*

(i) Consumers with open access permission by the proper commission should be provided with an interface meter; this meter interconnects the inter- or intra-state transmission system [4, 5].
(ii) The consumer who is connected to the distribution system and allowed open access should have the provision of interface meters as per directions of the proper commission.
(iii) The licensee connected to a central or state transmission utility should submit the location of interface meters in advance before the installation.

### *8.5.2 Consumer Meters*

- The licensee should install this meter either inside or outside the consumer premises.
- When the licensee installs the meter outside the consumer premises, he or she should provide a realtime display unit to indicate the electricity consumed by the consumer.
- The consumer meter reading should be taken into account, not the display unit, for billing purposes.
- When the commission allows, the direct electricity supply from the generation company to the consumers on a separate transmission system and the location of meters should be per the mutual agreement between them.

### 8.5.3 *Energy Accounting and Audit of Meters*

These meters are installed at the locations to facilitate the energy generated, transmitted, distributed, and consumed in the various parts of the power system and the energy loss [6]. The location of these meters should be as follows.

(i) Generating Stations:

- At the generator stator terminal.
- Generating station and unit auxiliary transformers high- and low-voltage sides.
- Feeders to various auxiliaries.

(ii) Transmission Stations:

- Energy accounting and audit of meters are installed for all incoming and outgoing feeders whenever interface meters do not exist.

(iii) Distribution System:

- Energy accounting and audit meters are installed for all incoming and outgoing feeders for the distribution system, when the system voltages are 11 kV and above.
- Depending upon the necessity for energy accounting and audit, the licensee may provide the meter on the primary or secondary side or both sides of the substation transformer including any distribution transformer.

## 8.6 Accuracy of Electromechanical Versus Electronic Energy Meters

Table 8.2 gives comparisons between meters.

## 8.7 Communication Facility in Meters

There are various types of communication facility available in energy meters. These ports enable easy, accurate, and automated meter reading. Nowadays meters having multiple communication ports are available [7, 8]. Some of the variety is as follows.

a. Optical communication
b. RS 232 communication
c. RS 485 communication
d. RJ 11 communication
e. Modem communication
f. Radio frequency (RF) communication

**Table 8.2** Comparison between electromechanical and electronic energy meters

| S. No. | Electromechanical energy meters | Electronic energy meters (TVM) |
|---|---|---|
| 1. | Energy measured is based on the principle of electromagnetic induction | Energy measured is based on the principle of sampling |
| 2. | Meter measures only active current (KWH) | Meter measures active, reactive, and apparent energies in all four quadrants |
| 3. | It has poor accuracy of measurement. It can measure up to class 1 of accuracy | It has better accuracy. It can measure up to 0.2 s class of accuracy |
| 4. | It does not have a data storage facility | It has a data storage facility and stores kWh, kVArh (lead and lag), and kVAh in forward and reverse directions with the help of energy accumulator (registers) |
| 5. | Meter measures cumulative active power only | Meter measures instantaneous phase voltages, phase powers, currents, frequency, phase sequence, rising demand, power factor, date, and time |
| 6. | Manual reading facility only | Data collection possible through the optically isolated serial interface, using a PC or MRI |
| 7. | No time of day (TOD) facility | Time of day (TOD) programmable |
| 8. | Maximum demand is recorded by means of a manual counter that can be reset manually | Maximum demand registers with programmable block/sliding methods, integration periods, and energy types |
| 9. | No load survey recorder facility | Load survey recorder facility available |
| 10. | No tamper-recording facility in such meters | Various tamper records available for detecting tamper/fraud |
| 11. | Unidirectional only. In case of change in phase and neutral direction of energy recording changes | Meter can be configured as a unidirectional/bidirectional meter |
| 12. | No auto reset facility | Different means of reset—automonthly, manual, or automonthly and manual reset |

Meters are read using a meter reading instrument (MRI), wireless communicator for RF meters, or through automatic meter reading (AMR) [9]. In a direct MRI downloading meter, the meter reader carrying a handheld device reaches each metering site, connects the optical port/RS Port/RJ port of the meter and downloads meter data as required. In modem-connected meters, the meter reader need not visit the metering location each time. Meter data can be downloaded from the central server located at the office by using the GSM/GPRS/CDMA/3G network. Today the central server has a scheduler using this meter scheduled to read at a particular date and time. Manual interference can be completely avoided using this methodology. In the RF communication method, the meter reader has to visit the metering site and search for nearby meters within the accessible range (normally 50–100 m). After finding the meter the same can be downloaded by establishing wireless communications [10].

## 8.8 Meter Calibration and Periodical Testing

### 8.8.1 *Interface Meter*

- The accuracy of each interface meter is tested using a standard reference meter by the owner at the site at the time of commissioning.
- Periodic testing is required to ensure accuracy, and also to avoid malfunctioning. As per the standard, meters are tested at least once in five years.
- The meters should be tested via a NABL-accredited mobile laboratory or by any accredited laboratory and recalibrated if meter records any abnormal or inconsistent electrical quantity as required by the manufacturer's works.
- Advance notice should be sent to the customer from the supplier to intimate the date and timing of testing; it is carried out in the presence of representatives from both sides.

### 8.8.2 *Consumer Meter*

- The accuracy of each consumer meter is tested using a standard reference meter by the owner at the site at the time of commissioning.
- Periodic testing is required to ensure accuracy, and also to avoid malfunctioning. As per the standard, meters are tested at least once in five years.
- The testing of the said consumer's meter above 650 V should cover the metering system including CTs, VTs, and the like. The meters may be tested through a NABL-accredited mobile laboratory or by any accredited laboratory and recalibrated if the meter records any abnormal or inconsistent electrical quantity as required at the manufacturer's works.
- Advance notice should be sent to the customer from the supplier to intimate the date and timing of testing; it is carried out in the presence of representatives from both sides.

### 8.8.3 *Meter Energy Accounting and Audit*

- The accuracy of each consumer meter is tested using a standard reference meter by the owner at the site at the time of commissioning.
- Periodic testing is required to ensure accuracy for both check meters and standby meters, and also to avoid malfunctioning. As per the standard, meters are tested at least once in five years without removal of CTs and PTs.
- The meters may be tested by a NABL-accredited mobile laboratory or by any accredited laboratory and recalibrated if the meter records abnormal or inconsistent electrical quantity using a measuring unit, secondary injection kit, phantom loading, and so on, as required at the manufacturer's works.

## 8.9 Best Practices for Meter Installation

To avoid erroneous meter readings, the following points should be taken care of while installing meters.

(i) Reversal of phase and neutral connection (single-phase).
(ii) Reversal of load connected on middle phase and earth (three-phase three-wire system).
(iii) Bypassing of neutral wire (three-phase four-wire meter).
(iv) Use of protection core of the meter.
(v) Influence of CT and PT wires.

- Long lead CT wires.
- Small size of the conductor.

(vi) For distribution transformer metering meter should be on LT side only.

- Meter specification should be of outdoor type; LT CT meters preferable.
- Best to locate so as to account for ATC loss.

(vii) For industrial and commercial metering ideal to locate near the front gate.
(viii) For agricultural supplies.

- Preferable on DT of each consumer.
- Meter should be located on the pole.
- Load restriction with built-in switches.

(ix) Outdoor type of meter mounting to be done inside enclosure.

## 8.10 Best Practices for Reliable Metering

(i) Proper stripping of cable/wire.

- Use of crimped pin socket.
- Use of chrome compound.
- Ensure proper contacts by lugging of cable/wire.
- Proper crimping of lugs.
- Ensure screws are tightened properly.
- Ensure proper antirusting bolt, nut, and spring washer.

(ii) Shake or rotate the cable to ensure proper seating of conductor strands in the terminal.
(iii) Terminal must completely cover the conducting (bare) part of the cable.
(iv) Meter should be properly mounted onto the panel/box and firmly fixed with screws.
(v) Ensure correct connection, that is, right polarity, phase association, and phase sequence.

(vi) Use of crimped joint and not twist joint when joining two wires.
(vii) Protection fuses must be used as per practice and appropriate places only.
(viii) Crossing of wire at meter terminals must not be allowed.
(ix) All wires must be numbered and ferruled for easy identification.
(x) No exposure of XLPE insulation.

- UV susceptible.
- Applying insulation tape is not the solution.
- Crutch point is highly vulnerable.
- Use of heat shrink breakout and sleeve.
- Embedding lug protection in sleeve.

(xi) Use of double-bolted long barrel lugs in LT side of distribution transformer.
(xii) Improper wiring of neutral wire for four-wire meters.
(xiii) Undersized neutral wire in three-phase loads.
(xiv) Adopting single earth wire return system in single-phase system.
(xv) Installing meter without confirming secondary CT rating.

## 8.11 Storage Methodology in Tri-vector Meters

### *8.11.1 Types of Tri-vector Meters*

Tri-vector meters have the following types of storage.

(i) Block
(ii) Sequential

(i) **Block Storage:** Separate memory will be allocated for each type of tampering for each phase. Tamper events will be recorded in this memory. The recording of each event is on a FIFO basis.

**Example**

The time of occurrence and duration of the tamper is recorded separately phasewise for each tamper for the latest 10 tampers. The number of tamper counts of each type of tamper is also recorded [11, 12].

If a tamper condition exists at the time of total power failure, then at the time of recovery of power, the tamper condition still continues without an increment in the tamper count. The power-off time is not added to the tamper condition.

Make 10 tampers of each type (R, VF), collect the data, and check for continuing tamper, tamper status, tamper count, tamper occurrence date and time, and tamper duration. Create one more event (R, VF); the first event will be erased and the second event will be shifted to the first event memory, the tenth event to the ninth event place and the new event will occupy the tenth event memory.

(ii) **Sequential Storage:** All tamper events will be recorded sequentially. There is no separate memory allocation for each type of tampering. Any one tamper can occupy complete memory if others do not exist, but in block a separate memory is allocated for each tamper.

**Example**

The time of occurrence and duration of the tamper is recorded separately phasewise for each tamper. The number of tamper counts of each type of tamper is also recorded. The tamper is recorded in the sequence of occurrence.

If the maximum tamper count is 100 events, make any 100 tampers, collect the data, and check for continuing tamper, tamper status, tamper count, tamper occurrence date and time, and tamper duration. Create one more event; the first event will be erased and the second event will be shifted to the first event memory, the 100th event to the 99th event place and the new event will occupy the 100th event memory (FIFO) [13, 14].

### 8.11.2 Storage Methodology

(1) **Snapshots:** Snapshots mean voltage, current, PF, kWh, and KVAH will be logged into memory at the time of occurrence or restoration of any type of tamper. Snapshot parameters vary from customer to customer.
(2) **Load survey snap:** Tamper and power fail snaps will be shown in the LS integration period if any tamper and power fail has occurred during that period. It can be checked on the spreadsheet as well as on a graph.
(3) **Total power fails:** Power fail records are recorded separately. The time of occurrence and duration of power fail is recorded for the latest 10 power failures. The power fail count is also recorded. Total power on duration is recorded as well.
(4) **Indications for voltage and current:**
    - 123 indication will be on LCD if phase voltage > 20% Vn
    - 1. R phase
    - 2. Y phase
    - 3. B phase
    - 123 will start blinking if phase current > 2% In
(5) **Persistence time:** Persistence time is the minimum time to identify the tamper with the tamper threshold by default occurrence; the time is 2 min and restoration time is 1 min.
(6) **Event:** Event means either occurrence or restoration of tamper.
(7) **Anomaly on display:** "*" (Star) will show on the LCD display after confirmation of any tamper and will clear after recovery of all tampers. The star will appear for any tamper occurrence.

## 8.12 Meter Testing

Energy meters are the heart of the distribution business. Revenue inflow to the power systems is guarded by energy meters. A distribution engineer is always behind accurate metering of the energy. There are various guidelines issued by regulators for periodic meter testing [15].

### *8.12.1 Preinstallation Laboratory Testing*

Manufacturers during supply of meters calibrate them within an acceptable range of accuracy. However, it is the distribution company's responsibility to check the meter accuracy before installation at the consumer's premises. In the process the distribution company checks the meter at its lab for the following tests.

- Accuracy testing by calculating the pulses or numbers of disc rotation at a particular load, standard voltage
- Mechanical inspection of terminal, body, and so on
- No-load, full-load tests
- Routine tests.

### *8.12.2 Onsite Tests*

Regulators have already defined the periodicity of a routine test for different kinds of meters. Obeying the same regulations, the testing engineer visits the metering site with all the metering testing devices. He or she inspects the site, sealing the condition of the meter in the presence of the customer/customer executive. The testing engineer tests the accuracy of the meter with a portable site meter testing device. That meter testing device has to be calibrated from a NABL-accredited laboratory.

After completion of the test, the testing engineer fills the meter inspection report and gets consent from the consumer's representative. In the case of any sign of theft/inaccuracy, the testing engineer opens the meter and seals it for *Panchanama* and invites the customer for joint inspection.

### *8.12.3 Joint Inspection of Meter*

Normally joint inspection of a meter is carried out if any tamper/misuse of the meter is detected. In this case the sealed meter is opened in front of the consumer, accessing authority, and manufacturing company's representative (if any). The meter is tested at a laboratory for accuracy calculation and witnessed by all. Where any tampering is detected the subsequent penalty is charged.

## 8.13 Meter Sealing

Electricity theft is a very big international problem in power utilities; nearly 20% of the electricity produced is stolen annually and creates big revenue losses. These costs are then usually imposed on honest consumers in the form of higher tariff rates. It is the responsibility of the distribution company and engineer to design a positive, reliable, security seal to enable checks on such theft. Specially designed seals are available on the market and specially researched designs enable the meters to be impossible to be tampered with without a trace. Once the said seal is broken it cannot be joined or returned to the meter [16]. The seal is designed for one-time use only and cannot be accidentally opened and must be deliberately cut by cutting pliers for its removal. The see-through body enables easy detection of internal tampering.

Following are some of the advanced variety of seals.

a. Met grip seals
b. Paper seals
c. Anchor security seal
d. Valve/meter cable seal
e. Tamper-evident security seal (with continuity check).

## 8.14 Latest Advanced Technology in Metering

### *8.14.1 Meter Data Acquisition (MDA)*

The meter data acquisition (MDA) system can acquire data from energy meters connected through an optical port/serial communication cord with all major meter manufacturers using GSM/GPRS/CDMA technologies. The meter data acquisition has these features:

- The data extraction can be granular (based on type of data, e.g., billing, instantaneous, load profile, etc.) and is incremental in order to minimize the data volume.
- The scheduling feature allows the user to configure the acquisition schedules for a location, set of meters, or for the office hierarchy such as division or subdivision.
- The system maintains a complete history of metering points and can compare datapoints of the same or different metering points.
- It supports report server integration.
- The system is scalable and integration with other applications of IT infrastructure.
- The solution is built on Java, XML, and Oracle technologies and has a browser-based Web interface for access.

### 8.14.2 AMR Technology

In India the following types of AMR techniques are being adopted.

- RS485 through Modbus RTU Open protocol
- PSTN (public switched telephone network)
- GSM (global system for mobile)
- CDMA (code division multiple access)
- RF (radio frequency)
- PLCC (powerline carrier communication).

### 8.14.3 Prepaid Metering

Figure 8.2 depicts prepaid metering.

- In order to reduce the gap between billing and collection efficiency a new methodology, prepaid metering, was introduced for educated customers.
- A proximity-based smart card prepaid energy metering system provides benefits and convenience to both utility and consumer.
- It reduces the billing cycle, increases revenue gain, and optimizes the use of energy in order to make the system more efficient.

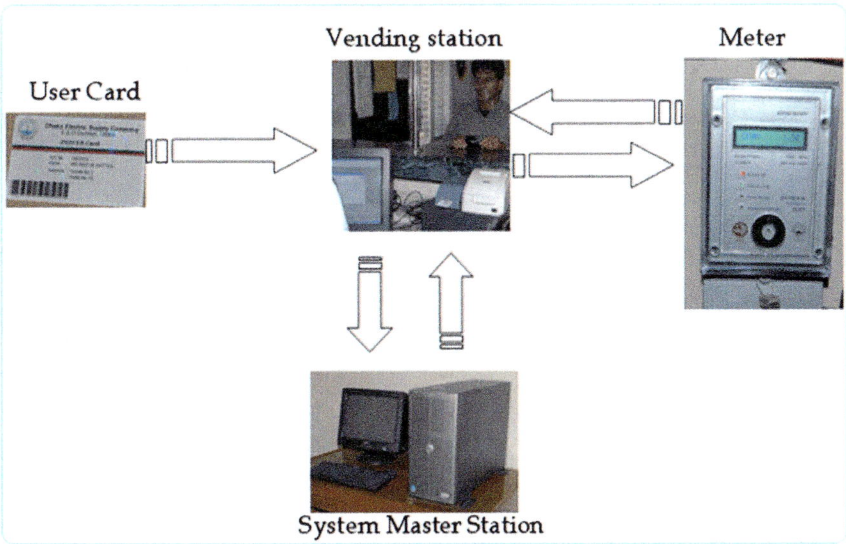


**Fig. 8.2** Prepaid metering

Various prepaid metering technologies are:

- Coin operated meters—obsolete technology
- Magnetic cards—obsolete technology
- Keypad meters
- Contact-based prepaid meters
- Proximity-based smart card.

## References

1. Orchard N (2000, Oct) The advance of metering. Utility Automation International, pp 31–32
2. Handbook for Electricity Metering (2002, Apr) 10th Edition. Edison Electric Institute, USA, pp 199–209
3. AMR/Metering Handbook (2006) The electricity forum. Ontario, Canada
4. Fryers T (2005) Metering international, issue 3, pp 84–85
5. IS Standards Related to metering: (i) IS: 13010, ac electricity meters induction type class 2,1,0.5. (ii) IS: 13779, ac electricity electronic meters class 2 and 1. (iii) IS: 14697, CT/PT operated ac electronic watthour meters class 2 and 1. (iv) IEC: 62053-11, ac electromechanical Kilo watthour meters class 2,1,0.5(v) IEC: ac electronic (static) kWh meter 60687 (class 0.2 s and 0.5 s) (vi) IEC: ac electronic (static) kWh meter 61036 (class 1 and 2) (vii) IEC: 61268, ac electronic reactive energy meters (class 1 and 2) (viii) IEC: 60044-1-2, Instrumentation Transformers (CT and PT)
6. Forst & Sullivan (2008) AMR market Report
7. Chris Bread and Logica (2008) Smart metering or dummies. Wiley, United Kingdom,
8. Central Electricity Authority (Installation and Operation of Static Meters) regulations (2006)
9. IEE conference publication No 462, Metering and Tariffs for Energy Supply, 25–28 May, 1999
10. Prepaid Electricity Meters: IEC 62055-41 and IEC 62055-51
11. Ripple control receivers for tariff and load control IEC 61037
12. Fryers T (2005) Metering internationals, issue 2, pp 84–85
13. IEC Standard 60044-7 Voltage Sensors (Electronic Voltage Transformers)
14. IEC Standard 60044-8 Current Sensors (Electronic Current Transformers)
15. Central Electricity Regulatory Commissions (2008) (Open Access in Inter-State Transmission) Regulations
16. Economic Growth in the Future (1976) Edison Electric Institute, McGraw Hill Book Company, New York, pp 257–286